Synthesis Lectures on Engineering, Science, and Technology

The focus of this series is general topics, and applications about, and for, engineers and scientists on a wide array of applications, methods and advances. Most titles cover subjects such as professional development, education, and study skills, as well as basic introductory undergraduate material and other topics appropriate for a broader and less technical audience.

Arijit Karmakar · Valentijn De Smedt ·
Paul Leroux

Integrated Time-Based Signal Processing Circuits for Harsh Radiation Environments

 Springer

Arijit Karmakar
IMEC-Ghent University
Ghent, Belgium

Valentijn De Smedt
KU Leuven
Geel, Belgium

Paul Leroux
KU Leuven
Geel, Belgium

ISSN 2690-0300   ISSN 2690-0327 (electronic)
Synthesis Lectures on Engineering, Science, and Technology
ISBN 978-3-031-40619-5   ISBN 978-3-031-40620-1 (eBook)
https://doi.org/10.1007/978-3-031-40620-1

This Springer imprint is published by the registered company Springer Nature Switzerland AG
The registered company address is: Gewerbestrasse 11, 6330 Cham, Switzerland

Paper in this product is recyclable.

Preface

This research primarily focuses on designing and implementing new architectures for integrated data converters based on time-based signal processing for critical reliability applications in harsh radiation environments. The radiation-hardened integrated circuits (ICs) were developed anticipating high radiation tolerance levels, particularly suited for high-energy physics (HEP) experiments, critical long-term space missions, and future nuclear-operated power plants. Mixed-signal interfaces like sensor readouts (resistive, capacitive, etc.) and clock interfaces are integral parts of various such critical applications, requiring assurance of reliable operation. In the presence of ionizing radiation, total ionizing dose (TID) effects arise due to long-term exposure, gradually changing the threshold voltage, charge carrier mobility, and leakage current of complementary metal-oxide-semiconductor (CMOS) transistors, which affects the performance of analog circuits. Single-event effects (SEEs) are instantaneous disturbances from charge deposition in the silicon (Si) when a charged particle strikes a sensitive node. In typical voltage- or current-based signal processing, the analog-to-digital converters (ADCs) comprise multiple voltage amplifiers, integrators, and comparators, the performance of which is significantly degraded by TID. Although TID effects can be minimized using smaller technology nodes (thin gate oxides), the impact of SEEs increases considerably. Time-based circuits designed in scaled CMOS technologies are relatively robust to TID. Also, in harsh radiation environments, time-based techniques make it easier to counter SEEs by employing various radiation-hardened-by-design (RHBD) techniques (majority voters, functional redundancy, C-element, etc.). The symbiosis between scaled technologies and RHBD techniques works to the advantage of time-based signal processing over voltage-domain signal processing.

As a part of this research, three CMOS-based IC prototypes were implemented in a commercial 65 nm CMOS technology and tested for performance validation and proposed working principles. Two different types of quadrature LC oscillators were implemented in the first project. The sensitivity of different performance parameters of the working prototypes with respect to TID was studied using X-ray irradiation (TID ≤ 100 Mrad (SiO$_2$)). Next, a radiation-hardened time-based $\Delta\Sigma$ capacitance-to-digital converter (CDC) was implemented based on MASH 1-0 configuration. The prototypes were validated experimentally in a radiation-less lab environment and with heavy-ion exposure (Xe-ion with LET 65 MeV.cm^2/mg). The CDCs could measure capacitance in the range of 0–3.75 pF with a clock of 100 MHz and have achieved ENOB of 12.9 bits with an energy efficiency of 0.18 pJ/conversion-step. Finally, in the third project, a differential $\Delta\Sigma$ time-to-digital converter (TDC) was implemented with a maximum T_{range} of 98(=±49) ns. It uses a time-based FIR filter in feedback to configure the first, second, and third orders of $\Delta\Sigma$-modulation. It consumes 11–15.8 μW from a 0.6 V supply and achieves the best state-of-the-art energy efficiency in the range of 15.6–89.2 fJ/conversion-step at 10 MHz clock frequency across multiple orders ($\leq$3) of $\Delta\Sigma$-modulation.

In addition, the research findings were valorized by collaborating with an industrial partner to develop a radiation-hardened frequency synthesizer for future space missions. During the research exploitation, several radiation-hardened on-chip integrated circuit components (a digitally-controlled oscillator, a multi-modulus divider, and output buffers) of an all-digital-phase-locked-loop (ADPLL) were implemented and characterized under ionizing radiation.

Leuven, Belgium   Arijit Karmakar
Geel, Belgium   Valentijn De Smedt
Geel, Belgium   Paul Leroux

Acknowledgements

We would like to thank a number of individuals as well as organizations for their significant contributions to the work in this book.

We would like to express our gratitude to Prof. dr. ir. Filip Tavernier (KU Leuven, Belgium), Prof. dr. ir. Greet Langie (KU Leuven, Belgium), Prof. dr. ir. Liesbet Van der Perre (KU Leuven, Belgium), Prof. dr. ir. Piet Wambacq (Vrije Universiteit Brussel-IMEC, Belgium), Prof. dr. ir. Johan Bauwelinck (Ghent University-IMEC, Belgium), and Dr. Ying Cao (Magics Technologies, Belgium) for all the engaging and inspiring discussions and their feedback on previous versions of this manuscript which helped us to refine and strengthen its content.

We would also like to acknowledge RADSAGA-ITN program under the European Union Horizon 2020 Research and Innovation Program with grant agreement number 721624 for the financial support of the research activities.

Contents

Abbreviations

ADC	Analog-to-Digital Converter
ADPLL	All Digital Phase-Locked Loop
BJT	Bipolar-Junction-Transistor
BOX	Buried Oxide
CDC	Capacitance-to-Digital Converter
CERN	European Organization for Nuclear Research
CLA	Clock Level Averaging
CMOS	Complementary Metal-Oxide-Semiconductor
COTS	Commercial off the Shelf
CSI	Current-Starved Inverter
DCC	Differential-Charge-Cancellation
DCO	Digitally Controlled Oscillator
DD	Displacement Damage
DEM	Dynamic Element Matching
DICE	Dual-Interlocked Storage-Cell
DMR	Dual Modular Redundancy
DTC	Digital-to-Time Converter
DUT	Device Under Test
DWA	Data-weighted Averaging
ELT	Enclosed Layout Transistor
ENOB	Effective Number of Bits
ESA	European Space Agency
FDSOI	Fully depleted Silicon on Insulator
FGMOS	Floating Gate MOSFET
FIR	Finite Impulse Response
FOM	Figure of Merit
FOXFET	Field Oxide Field-Effect Transistor
GRO	Gated-Ring-Oscillator
HDBL	Heavily-Doped-Buried-Layer
HEP	High-energy Physics

HIT	Heavy-ion Tolerant
IC	Integrated Circuit
IIR	Infinite Impulse Response
INL	Integral Non-linearity
ITN	Innovative Training Network
LDD	Lightly-Doped-Drain
LET	Linear-Energy-Transfer
LOCOS	Local-Oxidation-of-Silicon
LVCMOS	Low-Voltage Complementary Metal-Oxide-Semiconductor
LVDS	Low-Voltage Differential Signaling
MASH	Multistage-Noise-Shaping
MDLL	Multiplying Delay-Locked Loop
MMD	Multi-modulus Divider
MOM	Metal-Oxide-Metal
MOS	Metal-Oxide-Semiconductor
MOSFET	Metal-Oxide-Semiconductor Field-Effect Transistor
MSB	Most Significant Bit
NIEL	Non-ionizing Energy Loss
OTA	Operational Transconductance Amplifier
PCB	Printed Circuit Board
PCIe	Peripheral Component Interconnect Express
PLL	Phase-Locked Loop
PMT	Photomultiplier Tube
PQVCO	Parallel coupled QVCO
PWM	Pulse-Width Modulated
QEF	Quantization Error Filter
QVCO	Quadrature VCO
RADFET	Radiation-sensitive Field-Effect Transistor
RADSAGA	RADiation and Reliability Challenges for Electronics used in Space, Aviation, Ground, and Accelerators
ReRAM	Resistive Random Access Memory
RHBD	Radiation Hardened By Design
RHBP	Radiation Hardened By Process
RINCE	Radiation-Induced Narrow-Channel Effect
RISCE	Radiation-Induced Short-Channel Effect
RO	Ring Oscillator
SAR	Successive Approximation Register
SEB	Single-Event Burnout
SEE	Single-Event Effect
SEFI	Single-Event Functional Interrupt
SEGR	Single-Event Gate Rupture

SEL	Single-Event Latch-up
SET	Single-Event Transient
SEU	Single-Event Upset
SIFT	Software-Implemented Hardware Fault Tolerance
SMA	Sub-miniature Version A
SNR	Signal-to-Noise Ratio
SOD	Silicon-on-Diamond
SOI	Silicon-on-Insulator
SOS	Silicon-on-Sapphire
SQVCO	Super-harmonic coupled QVCO
SRAM	Static Random Access Memory
SRO	Switched-Ring-Oscillator
SSPM	Solid-state Photomultiplier.
STI	Shallow Trench Isolation
TA	Time-based Adder
TAU	Time-based Adder Unit
TD	Time-difference detector
TDC	Time-to-Digital Converter
TI	Time-based Integrator
TID	Total Ionizing Dose
TMR	Triple Modular Redundancy
TR	Time-based Register
TSPC	True Single-Phase Clock
TVC	Time-to-Voltage Converter
VCO	Voltage-Controlled Oscillator
VTC	Voltage-to-Time Converter

List of Figures

List of Tables

Introduction

1

Abstract

Ionizing radiation affects electronic circuits as well as living beings. Electronics constantly exposed to this kind of radiation eventually malfunction or produce inaccurate results. Critical reliability applications in harsh radiation environments like nuclear power plants, high-energy physics experiments, and space applications require integrated circuits (ICs), and those ICs need to be radiation-hardened to ensure operation without interruptions. Scientists and engineers in the fields of physics and medicine, as well as those working on nuclear power plants and particle accelerators, have spent years studying the consequences of radiation exposure and devising ways to lessen it. This introductory chapter briefly discusses the historical background of ionizing radiation and its effects. The research goal and objectives have also been introduced to the readers in this chapter.

1.1 Historical Introduction

The significant milestones of radiation physics discoveries closely followed the timeline of research on atomic physics. The initial breakthrough can be traced back to the discovery of X-rays by Wilhelm Röntgen in 1895. Formal radioactivity studies are considered to begin with the discovery of radioactivity by Henri Becquerel in 1901. The separation of radioactive elements by Marie Sklodowska-Curie and Pierre Curie in 1902 is a critical discovery in radiation physics. The initial investigation on γ-rays by Paul Villard at the beginning of the twentieth century was soon followed by the pioneering studies of α- and β-rays by Ernest Rutherford in 1907. By the first half of the twentieth century, many scientific observations and discoveries were made, which led to further progress in the research and development of the knowledge base in the radiation field [1]. In the 50s and 60s of last century, an increased interest was laid in developing compact nuclear reactors and building commercial nuclear power generators based on the outcomes and scientific discoveries of the Manhattan Project. Around that period, Enrico Fermi and Leo Szilard built "Chicago Pile-1," the first sub-critical research nuclear reactor at the University of Chicago.

So far, the research in this field has been primarily focused on understanding the mechanism of radioactivity and applying it in related applications. However, the aftermath of the Manhattan Project soon led the scientists to focus on the requirement of radiation effects studies. Systematic investigation of ionizing radiation's effects began with the advent of the "Space Age" in the later half of the twentieth century. The launch of Sputnik-1, the first artificial earth satellite, happened in 1957, followed by the discovery of Van Allen radiation belts around the earth by James van Allen in 1958. The "Starfish Prime" experiment in 1962 resulted in the failures of many on-orbit telecommunication satellites. Very soon in this time frame, the first on-orbit repair of a satellite was performed by thermal annealing and operating point adjustment to compensate for radiation damage. The ever-growing interest in space explorations was shortly complemented by the success of the Apollo moon landing and exploration program in 1969.

Earlier research on ionizing radiation effects primarily concentrated on understanding the mechanism of damage in bulk lattice structures, known as displacement damage (DD). As the bipolar-junction-transistors (BJTs) were the predominant devices used for electronics development in that time frame, the gain degradation due to the lattice damage caused by ionizing radiation was the central phenomenon that the researchers tried to focus on. However, the investigation into the failure of the Telstar-1 satellite in 1962 led to the discovery of surface-damage effects and the understanding of the dynamics of charge-trapping inside semiconductor oxide layers, which were later identified as total ionizing dose (TID) effects. Unlike BJTs, metal-oxide-semiconductor (MOS) devices were found to be majority carrier dominant, and that is why they believed to be intrinsically radiation hardened. Over time, as the research progressed, the observation of MOS's insensitivity toward ionizing radiation appeared incorrect. With technological advancement, the fabrication of MOS devices was made accessible, and the devices were used more frequently in place of BJTs for electronic circuit development. However, introducing MOS devices led to more complex failure mechanisms due to ionizing radiation.

The beginning of research around single-event effects (SEEs) in semiconductors (due to the random strike of charged particles) can be traced back to the observation of random anomalies in ground stations in the late 1960s [2]. With the development of digital logic circuits, this phenomenon became prominent and attracted the research community's interest. These random phenomena were later accounted for random memory upsets and were first formally reported in 1975 [3, 4]. Preliminary investigations about the recorded radiation events were found to be non-destructive "soft errors" in digital circuits. However, with time, as the researchers gained much more insights, several mechanisms were discovered, including latch-up and burn-out-like destructive events.

1.2 RADSAGA ITN Program

This research was facilitated as a part of the RADSAGA Innovative Training Network (ITN) Program [5], which was supported by Marie Skłodowska-Curie Actions in the European Union's Horizon 2020 research and innovation program. RADSAGA was the acronym attributed to RADiation and Reliability Challenges for Electronics used in Space, Aviation, Ground, and Accelerators. The ITN program was primarily targeted to initiate a collaborative network between industry experts, university researchers, and radiation test laboratories to facilitate research and innovation in radiation-affected electronics and accumulate knowledge for future endeavors.

The research, test, and training activities conducted in association with the RADSAGA ITN program involved four major inter-disciplinary work packages (Fig. 1.1) incorporating several researchers leading to innovations in radiation hardness, the study of radiation effects, and knowledge base of improved test methodologies. The focus of the first work package was on the integration and qualification of various radiation test facilities. The second work package primarily focused on establishing new methods of radiation study (e.g., aging) and design-level innovations (sensor interfaces and data converters) in the radiation hardness of integrated circuits used in various industrial applications. The third work package explores the system-level testing methodologies and guidelines for radiation hardness. The fourth work package was targeted to accumulate the learnings and the research outputs from the other three work packages. The research work conducted in this project was performed as a part of the third work package and primarily focused on establishing new radiation-hardened-by-design (RHBD) techniques for sensor interfaces and data converters implemented in commercial complementary metal-oxide-semiconductor (CMOS) technologies.

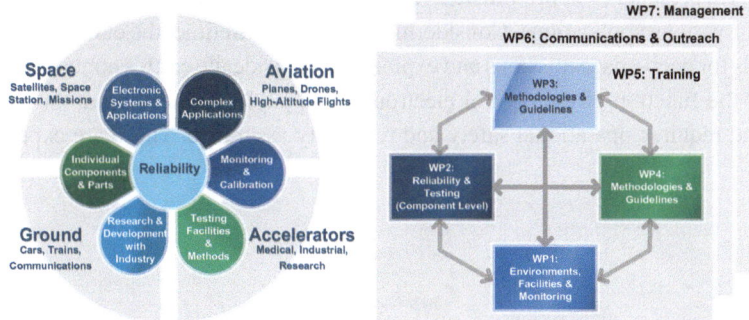

Fig. 1.1 Various instances of radiation-affected areas with reliability concerns (left) and a graphic overview of RADSAGA Work-packages (right)

1.3   Research Motivation

Ionizing radiation affects the physical properties of semiconductor materials in electronic devices and instruments. The effects are multi-fold and can be detrimental. Therefore, the radiation damages have been of utmost concern for the safe, reliable operation of various critical applications [6]. When asked about the various radiation-affected application areas, one would primarily consider medical, flights, nuclear power plants, space missions, and high-energy physics (HEP) experiments involving high-reliability concerns. In such scenarios, prolonged exposure to ionizing radiation could pose significant risks to the performance of electronic circuits, possibly leading to erroneous outcomes and functional disruptions. The experts on radiation effects and professionals like physicists, researchers, and nuclear reactor and accelerator engineers have long been working on understanding the theory and analyzing the effects of ionizing radiation. Over time, numerous studies [7–10] about the sources of radiation and the mechanism of radiation effects have essentially helped to model and characterize the radiation-related information and subsequently think about mitigation strategies.

Most sources causing ionizing radiation can be identified as γ-ray, X-ray, high- and low-energy protons, UV-ray, heavy ions, etc., originating from galactic cosmic rays, solar emissions, and radiation belts around planetary bodies [11]. Besides these, the α-particles emitted by the radioactive isotopes and impurities in chip packages could also potentially cause ionization inside the microelectronic components. Usually, a radiation event is registered when a highly energized particle strikes the surface of semiconductor devices. Such events either cause momentary voltage or current transients in the circuit nodes or lead to the accumulation of charges, affecting the device's properties. In the case of digital logic circuits, such transients might cause bit-flips, memory upsets, and functional interrupts. However, the performance of the electronic systems degrades throughout operation due to the cumulative dose, and eventually the system fails in extreme conditions.

Figure 1.2 presents an approximate overview of expected cumulative radiation dose levels [12] for various applications. Considering the mission lifetime, the cumulative radiation dose levels for deep-space missions and explorations outside the earth's atmosphere are less. However, the function of the control electronics and payloads installed in the satellites is crucial and requires operational safety and reliability assurance. Those are expected to be

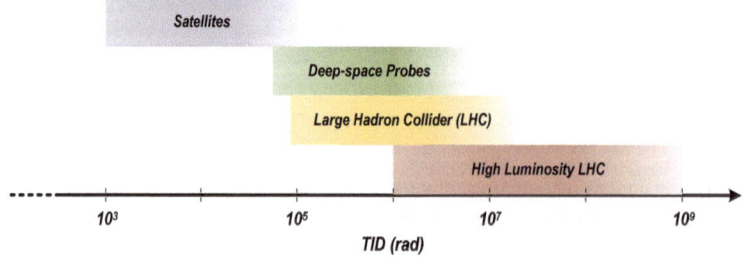

Fig. 1.2 Expected radiation dose levels for different applications (based on the data from [12])

protected against unwanted disruptions due to radiation strikes. In aviation, high-altitude flights used for earth remote sensing and monitoring are subjected to a comparatively mild flux of radiation particles. However, their complex electronic systems onboard still need to be safeguarded against the potential event of low-energy radiation strikes. At the ground level, the cloud-data servers, transport facilities, and communication interfaces are seldom subjected to soft errors due to ionizing radiation from natural sources and secondary emissions from the atmosphere. In the case of therapeutic medical centers, nuclear power plants, and particle accelerators for high-energy physics experiments, the artificially manufactured radiation levels could be extraordinarily high and pose a significant threat to the reliable operation and safety of the associated electronics and equipment.

1.4   Research Goal and Objectives

This research primarily focused on designing and implementing new architectures for integrated sensor readouts based on time-based signal processing for critical reliability applications in harsh radiation environments. In the presence of ionizing radiation, TID effects arise due to long-term exposure, gradually changing the threshold voltage, charged carrier mobility, and leakage current of CMOS devices. As a consequence, the performance of the electronic circuits is affected. SEEs are instantaneous disturbances from charge deposition in the silicon (Si) when a charged particle strikes a sensitive node. The primary research goals can be presented by three research questions: (1) How does ionizing radiation (TID and SEE) affect time-based signal processing? (2) What are the best ways of radiation hardening by design for integrated time-based sensor readout circuits in nanoscale technologies? (3) What are the design trade-offs for implementing low-power time-based sensor readout circuits with radiation tolerance?

In typical voltage- or current-based signal processing, the analog-to-digital converters (ADCs) include multiple voltage amplifiers, integrators, and comparators, the performances of which are significantly degraded by TID [13]. Although TID effects can be minimized using smaller technology nodes (thin gate-oxides), the impact of SEEs increases considerably [14]. Complex trade-offs between signal headroom, noise, linearity, bandwidth, power consumption, 1/f noise, and device matching introduce limitations to the performance of voltage-domain analog circuits, particularly data converters, by presenting significant design challenges. Compared to the widespread practice of voltage-domain signal processing, a time-based approach can be beneficial regarding technology scaling and achieving better tolerance to TID and SEEs. Time-domain circuits such as phase-locked loops (PLLs) and delay-locked loops (DLLs) are well-defined architectures, and the functional blocks are made of digital and semi-digital timing or clock systems that are relatively robust to TID. Also, in harsh radiation environments, time-based techniques make it easier to counter SEEs by employing RHBD techniques (majority voters, functional redundancy, C-element, etc.). The quantization capability (resolution) of time-to-digital converters (TDCs) is process technology dependent, and it enjoys a better scaling profile. However, the phase noise and increased

1/f noise corner still limit the realizable dynamic range in sub-100 nm technologies. Nevertheless, a distinctive feature of time-domain circuits is their substantial area compactness, and therefore hybrid architectures have become a recent trend in scaled processes.

Mixed-signal circuits like clock interfaces and sensor readouts (resistive, capacitive, etc.) are integral parts of radiation-affected areas such as high-energy physics (HEP) experiments, critical long-term space missions, and future nuclear-operated power plants. Various extensive studies [15–17] on the performance of LC-tank-based voltage-controlled oscillators (VCOs) operating under ionizing radiation are done, and several possible remedies are suggested with some design improvements. In comparison, the knowledge about the effects of radiation on the performance of quadrature voltage-controlled oscillators (QVCOs) is limited due to scarcity in the number of studies [18, 19] done on QVCOs under radiation exposure. Therefore, as a part of this research, the first goal was to experimentally perform a comparative study of TID-induced performance (oscillation frequency, phase noise, and power consumption) variation of different topologies of LC-tank based QVCOs. Following this, the next objective was the development of a radiation-hardened capacitive sensor readout suitable for application in dew point measurement during HEP experiments. Finally, the third objective was to implement a time-based data converter with optimized energy efficiency, which can be integrated within low-power frequency synthesizers for space applications.

A brief discussion about the rationale behind the choice of technology and signal processing techniques for radiation-hardened implementations has been provided in the following two sections.

1.5   Why the 65 nm CMOS Process Is Good?

Signals in mixed-signal interfaces require amplification and filtering and eventually get digitized by being converted from the analog-to-digital domain for further processing. The design of rad-hard mixed-signal interfaces (sensor interfaces, clock generators, etc.) needs mitigation strategies regarding cumulative dose and single-event effects. The significant challenges are to meet digital timing constraints (clock speed and delays) and analog performance criteria (gain, noise, and power) despite TID-induced degradation and to ensure reliable operation even in the presence of random SEEs. To achieve a circuit's tolerance against TID, one must first consider a suitable hardened technology in line with the specified requirements. The substrates in rad-hard technologies lessen the sensitivity of the striking ionized particles. Subsequently, the radiation hardness can be improved by combining it with various radiation-hardened design choices (rad-hard topologies, increased device sizing, calibration, etc.). The choice of technology depends on the trade-off between electrical performance, hardening capability, and manufacturing price. Two decades back, customized rad-hard processes were the preferable choice for the designers to implement rad-hard circuits due to the unavailability of radiation characterization data of commercial processes. One has to consider fabrication facilities with specially modified processes and bear high production costs. A rad-hard process usually needs to catch up several generations compared to commercially available leading feature sizes. The

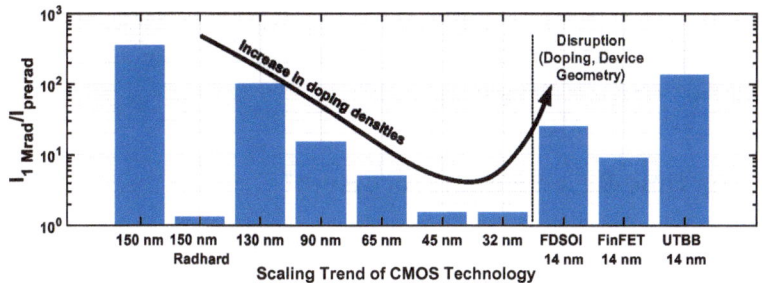

Fig. 1.3 Ratio of post-rad (1 Mrad(SiO$_2$)) to pre-rad leakage current with respect to CMOS technology (based on the data from [13])

designers were constrained with the choices and eventually had to compromise with the performance to achieve radiation tolerance.

However, as silicon technology evolved, the downscaling of feature size has made MOS devices intrinsically more immune to cumulative radiation dose effects [20]. As illustrated in Fig. 1.3, a decreasing trend of radiation-induced leakage current can be observed for different feature sizes of CMOS technologies [13, 21, 22]. The rate of radiation-induced hole trapping has decreased with the thinner gate oxide in newer technologies (sub-100 nm), and the generation of interface-trapped charges has also been reduced significantly due to electron tunneling [23, 24]. However, the trend of radiation-related performance improvement is disrupted with downscaling beyond 32 nm feature sizes for the technologies such as fully depleted Silicon on Insulator (FDSOI), Fin field-effect transistor (FinFET), and ultra-thin body and buried oxide (UTBB). As suggested in [13], the disruption could have been observed potentially due to differences in doping profiles and changes in device geometry. However, several recent radiation studies on FDSOI and UTBB technologies [25–27] have reported an increased degree of improvement in terms of radiation hardness. Nevertheless, prior to the 14 nm node, the trend of performance improvement with smaller feature sizes has made it possible for circuit designers to try mature, densely packed semiconductor processes and implement rad-hard designs with less power and higher clock speeds. In recent days, this has been complemented by the extensive effort of the radiation research community to provide radiation characterization data of mature commercial CMOS processes.

A commercial 65 nm CMOS technology has been chosen for the rad-hard designs and implementations during this research. Compared to more advanced technology nodes (e.g., 32, 28 nm), the fabrication cost of semiconductor IC prototypes in this particular technology node is much less. Therefore, a trade-off is taken into consideration between the performance and development costs. Several extensive studies [28–30] about various radiation-induced effects on the 65 nm CMOS process have already been conducted. According to the studies, this particular CMOS node does not showcase any novel radiation damage mechanism. It exhibits better radiation tolerance than its predecessors without significant compromise in analog performances. Undeniably, the 65 nm CMOS technology node has added benefits of increased logic density and decreased gate delays for digital logic implementations.

However, with these benefits comes an increased sensitivity toward radiation-induced SEEs. In the next section, further discussions have been provided on the mitigation strategies, mainly signal processing techniques.

1.6   Why Time-Based Signal Processing?

Besides the cumulative dose effects, the mixed-signal front ends must be safeguarded from radiation-induced SEEs to enable correct, reliable functioning [29]. As mentioned in the previous section, hardening electronic circuits against radiation can generally be accomplished by combining hardened processes and various circuit design techniques [6]. However, depending on the feature size of the rad-hard process used, the sensitivity of the various devices toward TID and SEE can appear contrasting and may diverge significantly. Thinner gate oxide in the scaled CMOS technologies helped the designers with intrinsic TID resilience but somewhat compounded the effects of SEEs. Following the scaling trend in CMOS technologies, the required critical charge for the logic levels to cause a bit-flip has been brought down significantly [31, 32]. Eventually, the reduced node capacitances in the smaller feature sizes and the reduced voltage headroom have contributed to the increase of probability of the occurrence of SEEs [20].

The mitigation of SEEs is usually performed at the signal processing stage by adopting various rad-hard circuit topologies, and redundancy techniques [1]. The radiation-induced transients could momentarily distort the signals in typical voltage- or current-based signal processing. As an effect, the signal quality is compromised, and the signal-to-noise ratio (SNR) is diminished. The strategies for alleviating such radiation-induced problems in the voltage or current domain are not straightforward. One should opt for low bandwidths for the circuit to limit the generation of the transients or introduce path delays to mask the propagation. However, in the digital domain, the mitigation strategies are more accessible to implement due to the availability of well-known RHBD techniques like triple-modular-redundancy (TMR), c-element, dual-interlocked cell (DICE), flipflops, etc. Furthermore, digital circuits have an inherent noise margin below which no transients could affect the circuit's operation.

A viable alternative for mixed-signal operation is time-based signal processing. This technique represents the sampled information as pulse width or time difference between two signal edges. Therefore, this method processes the signals using the algorithms and techniques developed for discrete-time signal processing. The circuits are made of pseudo-digital gates or blocks and therefore inherit the noise margin from the digital circuits. In addition, all the RHBD techniques developed for digital circuits can easily be adapted for time-based signal processing. Here the signals are represented by two logic levels, and the amplitude only depends on the transition edges. Unlike voltage or current domain, SEE-caused transients (within the noise margin) are less likely to introduce errors. Although for high-speed operation, the sensitivity toward single-event transients (SETs) increases, those can still be masked using digital logic-gate-based redundancy techniques (e.g., temporal and spatial redundancy). However, regarding the cumulative dose effects, there are no significant advantages of using time-based circuits instead of conventional voltage- or current-based circuits. While exposed to radiation over a period, performance variation is expected to happen for both types of signal processing, and there is no clear winner in this case. However, as mentioned in the previous section, several mitigation strategies can be employed to compensate for the radiation-caused damages.

One might be doubtful about what kind of compromises should be made in terms of performance while exploiting the added benefits of radiation tolerance. Microelectronic circuit designers have already explored time-based signal processing for various implementations (data converters, discrete-time filters, and sensor interfaces) of mixed-signal designs [33]. The trend of downscaling CMOS technologies is beneficial for time-based signal processing. Voltage-based implementations in smaller feature sizes suffer from performance degradation due to reduced voltage headroom, increased 1/f noise, and reduced dynamic range. Whereas for time-based, the reduced supply voltage does not significantly affect the circuit's performance, and the reduced gate delays improve the time resolution, eventually increasing dynamic range. Time-based circuits have no significant performance improvement at older technology nodes compared to voltage-based circuits. However, as investigated in [34], the performance of time-based circuits for sub-100 nm nodes is better than voltage-based counterparts in terms of dynamic range.

1.7 Organization of the Book

This book provides a concise overview of the fundamentals of ionizing radiation (sources, effects on semiconductors devices, radiation sensors, and mitigation methods). In addition, it introduces to the readers the details of the time-based signal processing technique in the context of ionizing radiation mitigation in integrated circuits. The chapter's content should make it easier for the readers to comprehend the various radiation-induced effects and recognize how critically important radiation hardness is. The research findings based on several IC prototypes substantiated with experimental data would assist the readers, particularly the IC designers in the radiation community to adopt their designs toward more robust, reliable operation in harsh radiation environments

In this introductory chapter, the historical background of ionizing radiation has been briefly discussed, and the research goal and objectives have been introduced to the readers. In addition to this current chapter, the book includes six more chapters describing the theory and implementation of prototypes and a concluding chapter.

Chapter 2 discusses the fundamentals of ionizing radiation while mentioning the sources and explaining, in short, the mechanism of interaction of the radiation particle with the semiconductor devices. The chapter also provides an overview of radiation-induced effects on devices, focusing on integrated circuits. Furthermore, it discusses the techniques and devices used for radiation measurement and mentions several circuit-level RHBD strategies for mitigating radiation-induced effects on electronics.

Chapter 3 discusses the fundamentals of time-based signal processing. It further provides an overview of various architectures of basic time-based building blocks and explains their working principles. At the end of the chapter, the pros and cons of time-based signal processing compared to traditional voltage-domain circuits and implementation-related challenges are discussed.

Chapter 4 presents the details of circuit-level implementation and radiation experiment of two different LC-tank-based quadrature oscillators. The results of the radiation study help to compare various performance parameters of the two types of oscillators and subsequently identify the vulnerabilities for future improvements. At first, the chapter discusses the prior art and the motivation behind the radiation assessment of the oscillators. Next, it provides the details of the architectures and elaborates on the mechanism of quadrature phase generation for both oscillators. In the end, the chapter elaborates on the details of the measurement setup under X-ray irradiation and presents the results obtained with detailed analysis.

Chapter 5 discusses the design and development of a radiation-hardened $\Delta\Sigma$-CDC. The implemented prototypes were validated experimentally in a radiation-less lab environment and the presence of heavy-ion exposure (Xe-ion). At first, the chapter discusses the prior art and then provides the details of circuit-level implementations while elaborating on the working principle of the critical building blocks. The chapter also mentions various RHBD strategies (foreground calibration, TMR, etc.) employed in the designed prototypes. Next, it presents the measured results with a detailed overview of the test setup used for the radiation experiment.

Chapter 6 discusses the design and development of the proposed $\Delta\Sigma$-TDC. The operation of the implemented prototypes is validated experimentally, and they demonstrate a single-loop multi-order (≤ 3) $\Delta\Sigma$-modulation. The proposed architecture uses time-based arithmetic units to construct a time-based FIR filter which is used in feedback to filter the quantization error resulting from the digitization. The filter coefficients are adjustable and can be configured to realize first, second, and third orders of $\Delta\Sigma$-modulation. At first, the chapter discusses the prior art and then explains the working principle of the proposed architecture. It also elaborates on the test setup for the TDC and presents the experimental results. This chapter further compares the performance parameters of the implemented $\Delta\Sigma$-TDC with respect to state-of-the-art designs.

Chapter 7 discusses the design and implementation of the radiation-hardened all-digital PLL (ADPLL)-based frequency synthesizer. This work was done as a part of research valorization in collaboration with an industrial partner. At first, the chapter provides the details of the architecture. It explains the circuit-level implementation of the ADPLL design featuring low-voltage differential signaling (LVDS), low-voltage CMOS (LVCMOS), and pulse-width modulation (PWM)-compatible outputs. Next, it presents the experiment setup for radiation assessment and the various performance metrics of the fabricated samples.

Chapter 8 concludes the book with an overview of the research activities performed as a part of this research study. It underlines the significant technical contributions and mentions some prospects for future research work. Furthermore, the chapter also discusses, in short, the details of the research valorization.

Fundamentals of Ionizing Radiation

2

Abstract

Ionizing radiation (γ-ray, X-ray, high-energy UV-ray, heavy ions, etc.) affects electronic circuits and living beings. It has been a significant concern for critical reliability applications like medical, aviation, space missions, and high-energy physics experiments considering safety and quality assurance. The radiation research community, including physicists, medical researchers, nuclear reactor, and accelerator engineers, has long been working on analyzing the effects of ionizing radiation. Over the past few decades, several studies about the sources of radiation and the physics of radiation effect mechanisms [7, 8] have helped the development of numerous radiation sensors. As silicon technology evolved, the downscaling of modern CMOS technologies has made MOS devices less sensitive to cumulative radiation dose effects [20]. As the gate oxide thickness has reduced, the rate of radiation-induced hole trapping has gone down. Also, the generation of interface trap charges has decreased notably due to electron tunneling [23], and subsequent neutralization of holes [24]. On the other hand, for scaled-down CMOS technologies, the critical charge required to cause a SEE has reduced significantly [31, 32]. As a result, the probability of SEEs in devices with smaller feature sizes has increased. Considering the wide variety of fields in the radiation environment, this chapter (Parts of this chapter were adapted from the published open-access article in MDPI, Radiation journal [6].) briefly discusses the fundamentals of ionizing radiation by mentioning the sources of ionizing radiation and briefly explaining the underlying mechanisms of various radiation effects. It also presents an overview of modern methods and devices developed for measuring the level of radiation and its effects on electronic instruments. Furthermore, it discusses various mitigation strategies for radiation-induced effects focusing on radiation-hardened-by-design (RHBD) techniques.

© The Author(s), under exclusive license to Springer Nature Switzerland AG 2024   11
A. Karmakar et al., *Integrated Time-Based Signal Processing Circuits for Harsh Radiation Environments*, Synthesis Lectures on Engineering, Science, and Technology, https://doi.org/10.1007/978-3-031-40620-1_2

2.1   Radiation Sources and Mechanisms

2.1.1   Sources of Radiation

Semiconductor materials inside electronic devices and instruments deployed in airspace, ground, and outer space encounter ionizing radiation every so often. The primary sources of radiation in the extra-terrestrial space environment [35, 36] are galactic cosmic rays (high-energy protons and heavy ions), intermittent solar emissions (low-energy protons, plasma, and magnetic flux), and radiation belts comprising charged particles accumulated around planetary bodies. Despite the shielding, the dose levels experienced by the microelectronics inside the devices are high and pose significant risks during space missions.

In comparison, the magnetic field of the Earth and its atmosphere generally safeguard the terrestrial environment [36, 37]. The effects of radiation from extra-terrestrial sources are reduced mainly due to the interaction with the Earth's atmosphere. Therefore, it encounters radiation primarily from the high-energy flux of cosmic-ray neutrons generated in nuclear interactions between the protons and the atmospheric elements. Another significant source is the α-particle emitted from the natural radioactive isotopes and local impurities in electronic devices (like chip packages). These radiation sources mainly account for electronic products' functional interrupts and reliability failures.

On the contrary, dose levels are very high as encountered by the electronics in artificial radiation environments that exist in numerous applications like nuclear energy plants, medical and high-energy physics experiments [38–40]. The primary radiation sources are X-ray, γ-ray, heavy ions, and high-energy protons. Even though these sources can account for remarkably high-dose rate particularly affecting the microelectronics, the total amount of absorbed dose by the personnel is kept within manageable limits due to distancing from the radiation sources and short exposure time as mandated by the safety standards.

2.1.2   Radiation Mechanism

As mentioned earlier, the sources of radiation (high-energy photons from X-rays and γ-rays, neutrons, charged particles: β, proton, α, and heavy ions) are abundant, and most of them play vital roles in affecting the operation of microelectronics components. The impact of radiation can be broadly categorized into three effects: (a) TID, (b) SEE, and (c) DD. The first two are the predominant effects of ionizing radiation. The last one is the outcome of non-ionizing interaction with high-energy particles, also referred to as non-ionizing energy loss (NIEL).

Depending on the technology used, the device sensitivity to TID and SEE can be contrary to each other and may differ significantly. Modern CMOS technologies favored TID resilience with thinner gate oxides. However, unlike TID, it has not helped the mitigation of SEEs. The concurrent effects of decreased power supply and reduced node capacitance in scaled technologies have worsened the issue.

Total Ionizing Dose (TID)

A radiation event is triggered by the incidence of a highly energetic particle. Most of the energy lost along the interaction path results in the excitation of electrons, henceforth creating charged electron-hole pairs as denoted by e^- and h^+ in Fig. 2.1. In conductors and semiconductors, the excess electron-hole pairs generated due to the interaction will eventually recombine or mobilize by drift and diffusion. In comparison, the outcome differs in the case of insulators like the most common SiO_2 present in CMOS technologies. Due to the difference in mobility, the excess electrons are swept away promptly by diffusion and drift (in the presence of an electric field). Conversely, the holes migrate comparatively slowly by "hopping" toward the $Si - SiO_2$ interface through localized shallow traps in the bulk silicon. As illustrated in Fig. 2.1a, these processes either lead to the formation of charge due to the hole trapping at a bulk defect or the generation of interface traps due to interaction with hydrogen ions (H^+). During irradiation, the trapped charges account for the cumulative TID effect in MOS field-effect transistor (MOSFET) structures [41]. Over time, the effects of accumulated positive charges manifest in terms of a change in threshold voltage, driving strength, leakage current, and increased flicker noise [42], thus deteriorating the device behavior. The TID-induced performance degradations can be attributed to radiation damages mainly occurring in three different regions of semiconductor devices. The first one happens in the gate oxide, potentially resulting in threshold voltage shifts. The second type of damage accumulates in the channel edges, resulting in the parasitic edge transistors turning on. The third kind appears in the isolation oxide, which can increase inter-device leakage current. The effect of TID is different in NMOS compared to PMOS devices. The interface and oxide-trapped charges are both positive in the case of PMOS devices. Therefore a negative threshold voltage shift is observed in such devices. In the case of NMOS devices, the

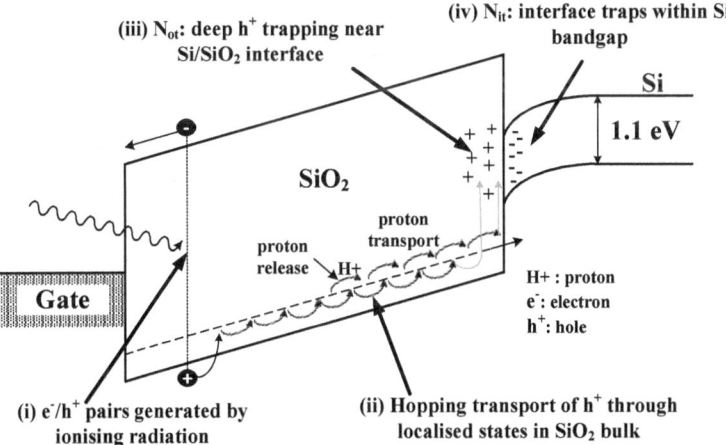

Fig. 2.1 Energy band diagram of a positive gate-biased MOSFET device showing the effect of ionizing radiation on carrier generation, transporting, and trapping

interface and oxide-trapped charges are opposite. As a result, a positive threshold voltage shift is observed, but the change is not as much as in PMOS devices. Over time if exposed to radiation, the NMOS devices become leakier and appear difficult to turn off, whereas the PMOS devices appear difficult to turn on. A portion of TID-induced trapped holes in the oxide region is eliminated by tunneling at ambient temperature or thermal recombination at elevated temperatures during annealing. The BJTs exhibit an increased sensitivity to TID-induced degradations at modest dose rates, whereas the majority carrier devices, such as MOSFETs, do not. At higher radiation dose levels, particularly in the case of BJTs, electrons are not taken away as quickly as they are at low radiation dose rates. Consequently, some oxide-trapped holes recombine with surplus available electrons, reducing the impact of TID-induced degradation. In addition, the cloud formation of charged particles in the oxide region reduces the number of interface-trapped charges by inhibiting proton release and its transport to interface regions.

Single-Event Effects (SEEs)

When the striking, energetic particle hits a semiconductor device, for instance, in the proximity of a doped region (drain or source junction of MOSFET devices), it induces an excess number of electron-hole pairs. At the onset, the ionizing interaction with the radiation produces a funnel-shaped path of free carriers [43] as shown in Fig. 2.2. If the device is under bias, the transport of carriers prevails over their dissipation through recombination. Most of the charge carriers drift toward the opposite polarity bias. The rest is collected through slow diffusion. The inevitable influx of charged carriers constructs a current spike, potentially resulting in transients in the electronic circuit. The various effects caused by the instantaneous disturbances triggered by the bombardment of charged particles are collectively referred to as single-event effects (SEEs). Whenever an energetic ion traverses through active devices, the archetype of all SEEs, a single-event transient (SET), occurs. In circuits with combinational logic or analog components with no memory or latches, SETs cause momentary disruptions in the operation. The circuit functionality eventually returns after a short duration once the excess charge has been swept away. However, in digital sequential circuits with memory, the transients may change the data state of the affected node, and such erroneous states can cause systematic failures. These effects are commonly referred to as single-event upset (SEU) in memory or a single-event functional interrupt (SEFI). Typically these effects (SETs, SEUs, and SEFIs) are non-destructive and do not damage the devices and, therefore, are collectively referred to as soft errors. The errors due to the SEEs can also be permanent (hard errors) and catastrophic in the case of single-event latch-up (SEL), single-event gate rupture (SEGR), and single-event burnout (SEB). SEL happens when the transient current activates the parasitic thyristor (p-n-p-n) and short circuits the power to ground, resulting in a high inrush of current through the device. Generally, an SEL may or may not cause permanent device damage but requires a power cycling of the device to resume normal device operations. SEGR happens when the radiation strike of heavy ion

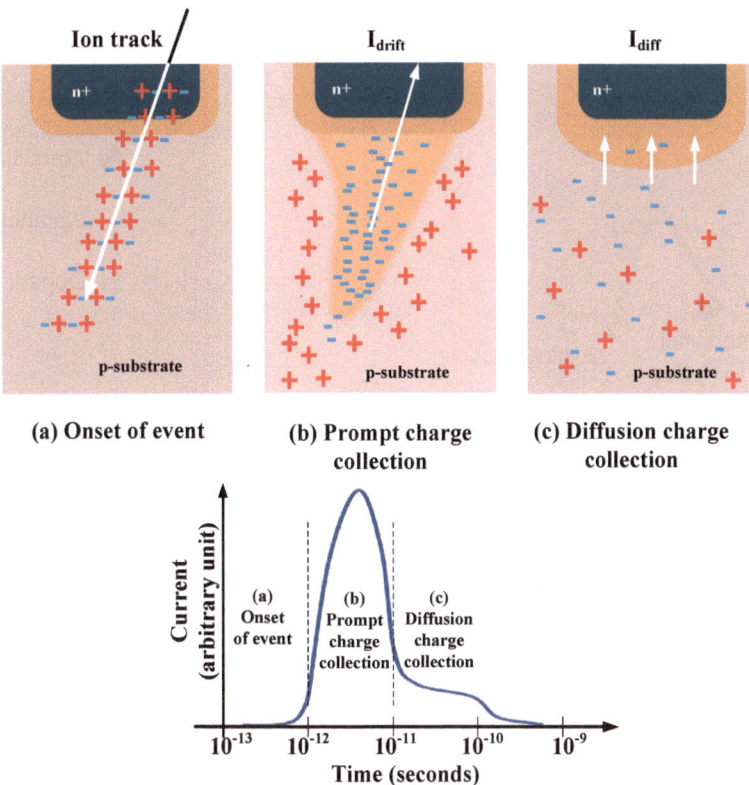

Fig. 2.2 Illustration of radiation-induced processes in a reverse-biased n+/p diode and the resulting current transient caused by the radiation strike

results in a breakdown of gate oxide and subsequent creation of conducting path through the gate oxide of a MOSFET. SEB is most prominent in power MOSFETs and can cause an abrupt and catastrophic breakdown in the device structure. The passage of the highly energetic striking heavy ion locally generates an excess amount of charge, which turns on a parasitic n-p-n transistor intrinsic to the power MOSFET. After that, the high amounts of current and voltage in the device cause a second breakdown of the parasitic BJT, ultimately melting the device.

Displacement Damage (DD)

Apart from the previous two phenomena (TID and SEE) causing excess charge generation, the radiation can also cause physical impairment in the crystal structure of the semiconductor material via NIEL. This effect is often called displacement damage (DD) [44, 45], and the primary sources are energetic neutrons, protons, or electrons. Secondary electrons

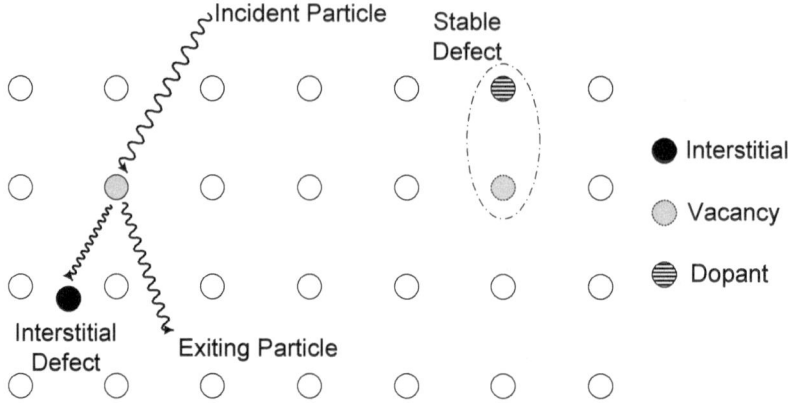

Fig. 2.3 Lattice defect clusters produced by a single ion or particle

generated from the interaction with γ-rays and high-energy X-rays can also cause DD if they have sufficient kinetic energy. Unlike the ionizing effects (TID and SEE), DD creates prolonged damages which are challenging to anneal out. The following are the fundamental processes (Fig. 2.3) that contribute to the degradation of materials and electronics in a radiation environment that results in displacement damage: (i) during irradiation, the Frenkel defects are created consisting of an atom vacancy or interstitial in the silicon lattice; (ii) these defects generate deep or mid-band traps which increase the thermal carrier recombination and generation and thus affect the free carrier density; and (iii) over time, the accumulation of a large number of defects degrades the semiconductor device's electrical and optical properties. Numerous factors, such as particle type and energy, irradiation temperature, material impurity type (i.e., n- or p-type), and concentration, affect the efficacy of damage caused by DD. DD-induced recombination centers cause the carrier lifetime to decrease, which in turn causes gain degradation in BJTs. Trapping the charged carriers in defect centers also increases the transfer inefficiency of charge-coupled optical devices [46]. In comparison to BJTs and optical devices, MOSFETs are robust in environments involving DD doses. MOSFETs are majority carrier surface devices with much higher carrier densities under normal operating conditions. The cross section of the thin channel region through which the charges transport is exceedingly small. Therefore, a significantly higher number of defects are needed to affect MOSFET characteristics. Once the incident energetic particles produce a mixture of isolated and clustered defects in the device's lattice structure, the defects eventually try to orient themselves to form more stable configurations. Typically as a part of the forward annealing process, the amount of radiation damage and its effectiveness is reduced. In contrast to the typical process, sometimes reconfiguring defects with time at increased temperature may result in more effective defects, often referred to as the reverse annealing process.

2.2   Radiation Measurement Techniques

The effects of ionizing radiation on electronic circuits and the living human body are extremely critical and multi-fold and have been a major concern for various critical applications. Continuous exposure to radiation affects the performance of electronic systems by producing erroneous results and leading to functional failure. It requires accurate estimation of the radiation dose-related information (dose rate, accumulated dose) to assess the system performance (known as radiation dosimetry) and then look for corrective measures. In the early 60s and 70s [47], the first set of basic radiation sensors was developed based on lead (plumbum) zirconium titanate ferroelectric materials [48], strain gauges [49], CaF_2 thermal luminescent dosimeters [50], Si calorimeters [51], etc. These radiation sensors were mainly used to measure absorbed dose and dose rate-related information. MOSFET devices were first identified to record the absorbed dose and applied as a radiation sensor for space environment in [52]. The phenomenon of SEE, comprising the "funneling" effect in silicon, was discovered and experimentally illustrated in [9, 10]. Following this, the quest for space explorations and satellite development focused on implementing another type of radiation sensor, namely, the particle detector, on identifying the radiation-induced faults in electronic systems. Over time, the increased understanding of radiation interactions [53, 54] and energy deposition processes has helped extend the sensors with improved response and new design techniques. From the early 80s, the theory of micro-dosimetry, particularly dealing with very low-dose-level effects on microelectronics, was explored and studied with an emphasis on radiation oncology applications [55]. In the last few decades, numerous techniques based on semiconductor devices, such as diodes, MOSFETs, and solid-state photomultiplier (SSPM), have been reported to estimate the absorbed dose of radiation with sensitivity varying several orders of magnitude from μGy to MGy. In addition, mitigating soft errors in integrated circuits requires the detection of charged particle-induced transients and digital bit-flips in storage elements. Depending on the particle energies, flux, and application requirements, several sensing solutions, such as diodes, static-random-access-memory (SRAM), and NAND flash, are reported in the literature.

The radiation events as studied are considered to be randomly distributed over irradiation time following Poisson statistics. Depending on the interaction of semiconductor materials with radiation, the sensors in active mode may respond to an individual quantum of radiation (particle detection) or collective interaction with multiple quanta of radiation over a period (dose measurement). However, in the passive mode of operation, when there is no external energy source connected to the semiconductor devices and associated circuits, the energy deposited on the devices by individual particle strikes accumulates over time. Later, when the circuits are biased and the sensors are measured, the accumulated energy only estimates the radiation dose. As a result, it fails to identify and measure the energy deposited by individual radiation events. In the active mode of operation, the radiation sensors and associated circuits are biased using an external energy source and support individual particle detection and dose measurement. In active mode, the sensors can operate either in pulse counting or integral mode. In the case of Si-diode- or Si-photodiode-based detectors, the readout electronics monitor the radiation-induced bursts of currents. The charge collected, the time integral of the current throughout the detection (integral method), provides a measure for the energy deposited. The energy information is used in radiation spectroscopy to estimate the energy spectra. Over a period, the integrated charge-related data is used to indirectly measure the accumulated dose. In pulse counting mode, the charge collected is compared to a predefined threshold and causes an event to the counter for adding up to the particle detection. In the case of memory-based sensors, such as SRAM, the critical charge required for a bit-flip in memory elements acts as a threshold for particle detection. The readout method used in 3D NAND-based flash memories presents multiple options for threshold comparison, allowing an estimation of energy spectra together with particle detection. In the case of individual MOSFET-based sensors, the real-time collection of charge generated from radiation is inconvenient and therefore has limited use as particle detectors. These sensor elements act as integrating devices and employ indirect techniques to estimate the total dose information.

The sensors, when operating in integral mode, are usually referred to as radiation dosimeters which help in the evaluation (directly or indirectly) or measurement of the cumulative radiation quantities (TID, kerma, dose rate, and exposure time) [56]. The integration of information can be done by the readout electronics (Si-diodes) or by the sensing device themselves (MOS-based sensors). The readout method follows different techniques such as threshold voltage V_{TH} measurement, current or voltage reference sensing, and time-based (frequency) output. It can be categorized based on the sensing device used, such as silicon diodes, SSPM, fully depleted silicon-on-insulator (FDSOI) varactors, and MOSFET devices such as radiation-sensitive field-effect transistors (RADFETs), field-oxide field-effect transistors (FOXFETs), and floating gate MOS (FGMOS). Radiation sensors, called particle detectors when operated in pulse counting mode, help identify radiation events, particularly soft errors in integrated circuits. The methods used are predominantly current, voltage sensing, or memory read operations in digital memory structures.

A more detailed individual overview of the various semiconductor-based radiation sensors is given in the following paragraphs.

Silicon Diode-Based Radiation Sensors

This method uses a diode's reverse-biased p-n junction for dosimetry measurement. The radiation-induced excess charges are collected with an external electric field and amplified using a charge-sensitive pre-amplifier. The resulting current or voltage transient from the amplified charge is compared with a threshold and shaped into a pulse to cause a trigger to the digital counter. Thus, while operating in pulse counting mode, the Si-diode-based radiation sensors are used as particle detectors. Alternatively, in the integral mode of operation, a readout circuit measures the transient voltage or current. The time integral of the measured quantity provides an estimate of the energy deposited during a radiation strike, which indirectly measures the total accumulated dose on the sensing devices. Si-diode-based dosimeters like single-sided silicon strip detectors [57] and active pixel detectors [58] are moderately sensitive and suitable for high-dose radiation measurement as present in pulsed radiation fields (X-rays, electron linear accelerators) [56]. Over time, the sensitivity of the diodes reduces due to the damage caused by radiation and needs to be re-calibrated to perform accurate reliable measurements. In addition to particle detection and dose measurement, the intensity of the current transient provides an estimate of the energy spectra and linear-energy-transfer (LET) of the striking high-energy particle.

Silicon Photodiode-Based Radiation Sensors

This type of radiation sensor uses Si-photodiodes as the sensing element. Four types of photodiodes can be used for radiation measurement [59]: the P-N junction photodiode [60], the P-I-N (p-type, intrinsic, n-type) photodiode [61], the avalanche photodiode [62], and the Schottky photodiode [63]. Compared to standard P-N junction photodiodes, P-I-N photodiodes provide better sensitivity as the vast intrinsic region helps achieve high charge generation per radiation strike. The highest sensitivity to radiation among photodiodes is usually obtained in Avalanche photodiodes due to the Avalanche breakdown action. However, Avalanche photodiode-based detectors need larger bias voltage and produce higher noise levels. Schottky photodiode-based radiation detectors are based on compound semiconductor structures (SiC) and provide a fast response time because of low operational capacitance.

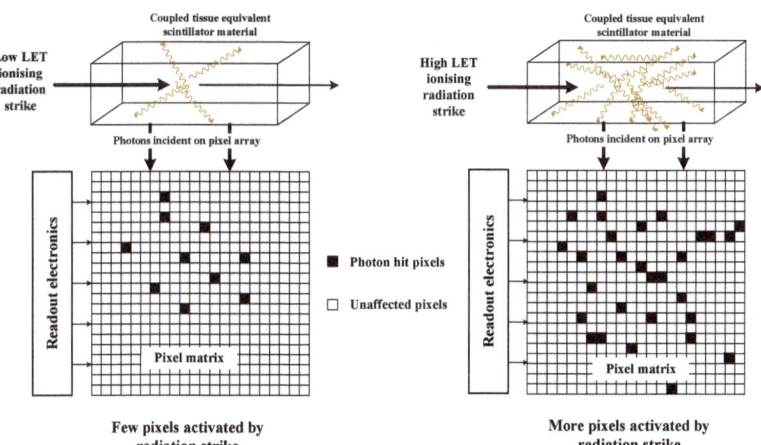

Fig. 2.4 Illustration of the working principle of a scintillator coupled SSPM-based dosimeter

The photodiodes integrated with SSPMs are alternatives for vacuum-based photomultiplier tubes (PMTs). An illustration of the CMOS SSPM chip coupled with scintillation material is provided in Fig. 2.4. When an energized radiation particle impinges on the dosimeter surface, the scintillator absorbs the energy and re-emits it as light. The generated photons hit the pixels in the SSPM matrix, and the device produces a current transient detected by the readout interface to estimate the accumulated dose accurately.

Radiation-Sensitive MOSFET Devices

The mobility and the threshold voltage (V_{TH}) of the MOSFET devices experience changes due to the radiation strike from high-energy particles. Initially, the MOSFET-based dosimeters were used in space [52] where the PMOS transistors were used in diode configuration to track and measure radiation dose through observing the V_{TH} shift. The RADFETs were provided with very thick gate oxide to increase the radiation sensitivity, which was used in nuclear facilities [64]. The devices were also used for radiation dosimetry, medical diagnosis, and radiotherapy treatments [65]. The fabrication of RADFET devices requires an ad hoc process, which makes it challenging to implement them with commercial CMOS technologies evolving to thinner oxides. To overcome the lower sensitivity of the RADFET devices, FOXFETs (Fig. 2.5) are developed where the field oxide (used as the passivation layer to isolate CMOS circuits) is utilized instead of the gate oxides.

Alternatively, FGMOS devices have also been explored for various dosimetry applications [66, 67]. The floating gate structures can be implemented by connecting the poly-silicon gate of a MOS capacitor to the gate of a sensing MOSFET device in standard CMOS technology [68] as shown in Fig. 2.6a. Otherwise, it can also be implemented by using an additional insulated poly-silicon gate in a custom two poly-CMOS process [69] as shown in Fig. 2.6b. When exposed to ionizing radiation, charges accumulate in the oxide region and neutralize the initial charge on the floating gate. This results in a change in V_{TH} [66], and the readout circuit integrated on-chip with the sensing devices measures the V_{TH} shift and provides a

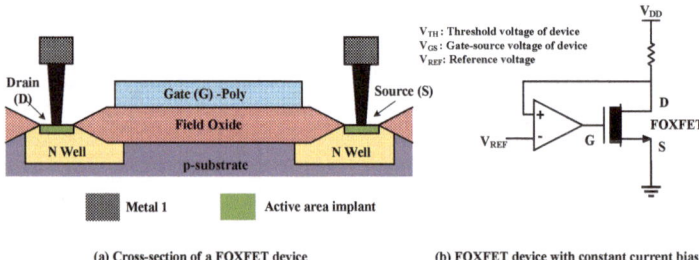

(a) Cross-section of a FOXFET device   (b) FOXFET device with constant current bias

Fig. 2.5 **a** Cross section of a FOXFET device used in dosimeter and **b** readout circuit for threshold voltage measurement of a FOXFET-based dosimeter

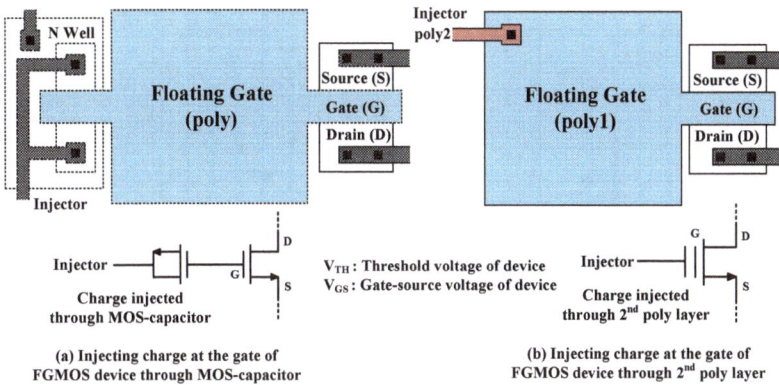

Fig. 2.6 **a** Top view of the FGMOS device integrated with MOS capacitor for charge injection into gate oxide and **b** top view of the FGMOS device with an additional poly-layer used for charge injection

proportional current [70] or voltage output [71]. Alternatively, the FGMOS devices are used to implement current-starved inverters and provide a frequency output, the change of which is dependent on the total accumulated dose [69]. FGMOS-based dosimeters are typically very sensitive to radiation dose, and therefore the sensor responses get saturated and suffer from non-linearity for dose levels more than 100 Gy. These provide high resolution (500 mrad [69]), but the application of these sensors is only limited to low-dose level (10–100 Gy) measurement. Compared to RADFET and FOXFET devices, FGMOS-based radiation sensors are recently preferred, considering the ease of on-chip integration with the peripheral circuits.

SRAM-Based Radiation Monitor

In the presence of radiation hazards, SEUs can easily affect the circuits that store data, for example, flipflops, RAM, and storage circuits. Among these, SRAM (static random-access memory) is the most typical circuit to monitor SEUs in electronic systems. It comprises two cross-coupled inverters, which create a stable bi-states structure, as shown in Fig. 2.7. The principle of SRAM-based particle detectors is simply searching the memory for bit-flips. These can be single-cell upsets or multi-cell upsets. As technologies have scaled down, single particle strikes can upset multiple cells simultaneously, depending on the particle energy. As indicated in [72], the ratio of single- and multi-cell upsets can be used to estimate the particle energy. The radiation detectors used by the European Space Agency (ESA) [73], the European Organization for Nuclear Research (CERN) [74], and satellite mission [75] have embedded multiple commercial off-the-shelf (COTS) SRAMs for SEU monitoring. The first purpose of these circuits is to monitor the radiation environment in space and ground

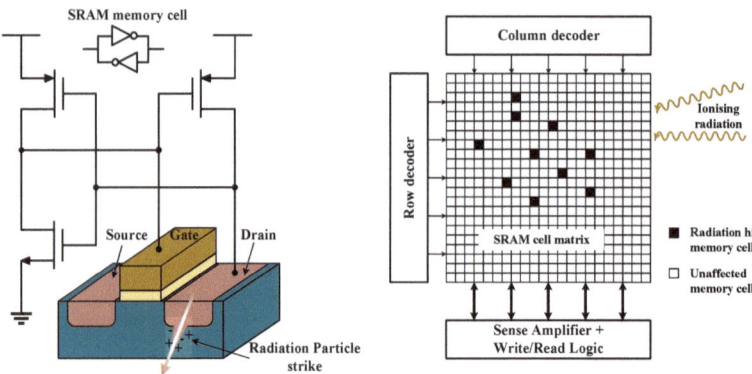

Fig. 2.7 Mechanism of radiation strike and memory bit-flips on SRAM nodes

facilities to guarantee that the environment's radiation intensity is in line with the design specification of the system and that the data is reliable. The second purpose is to monitor the flux, beam profile, and beam homogeneity at particle accelerator facilities such as RADEF and CHARM [73]. SRAMs fabricated in smaller technology nodes (65 nm and less) can achieve a higher SEU sensitivity and resolution. Most COTS SRAMs have an embedded error-correcting-code feature, which can correct some SEUs complicating or obstructing their use as radiation monitor [76]. Additionally, when testing for MCUs, one needs to be aware that the physical order of the cell array may differ from the logical (address) order through scrambling. The susceptibility of COTS SRAM circuits to SEL-induced failures is higher than custom-designed SRAM sensors. The COTS SRAM circuits are typically aimed to achieve high density. Therefore, the wells and doping areas are closer than usual chips, making them more prone to latch-up-induced circuit failures.

3D NAND Flash-Based Radiation Monitors

FGMOS devices are highly effective in dosimetry applications to measure the amount of TID in a radiation environment [77]. Recently, the floating-gate-based flash memories appeared as a viable alternative [78] to be utilized as particle detectors. As the ionizing particle impinges on the devices in flash memories, the excess charge impacts the floating-gate-based storage element. Subsequently, it causes a threshold voltage shift in the FGMOS devices inside the memory cell. Nowadays, the readout circuits inside the flash memory cells can identify the radiation event from the change in threshold voltage and collectively measure the number of upsets like SRAM-based radiation monitors. These particle detectors can also estimate the charge deposited due to the interaction with radiation by accurately measuring the threshold voltage shifts [79]. Unlike the SRAM-based solutions, it is feasible to characterize the LET of the striking particle in flash memories and extract information about the nature of the radiation (heavy ion, proton [80], neutron [81]). Although the 3D NAND flash-based particle detectors are beneficial in terms of efficiency and added features, the complexity of integration with planar CMOS technologies presents a difficulty in implementation. In addition, the device noise affects the accuracy of estimation of energy absorbed in radiation requiring complex readout schemes targeting low-noise mixed-signal readout implementations [82]. Another notable drawback is the long readout time needed to process a large number of memory cells, leading to increased power expenditure.

Variable Capacitor-Based Radiation Sensor

Variable capacitors or varactors implemented in CMOS technologies (bulk-CMOS and silicon-on-insulator) are viable alternatives to the diode and MOSFET-based solutions used for radiation sensing. Like these solid-state dosimeters, the capacitor-based sensors measure the accumulated dose depending on the ionizing effects inside the dielectric material. Although MOSFET-based dosimeters are very popular regarding reliability and added features, their DC operation methods are susceptible to various disturbances, particularly thermal and flicker noise sources [42]. In comparison, variable capacitor-based dosimeters can overcome the noise limits and accomplish ultra-high sensitivity by employing AC measurement techniques [83, 84]. Under radiation exposure, the capacitance-voltage (C-V) characteristics change due to trapped charges near the $Si - SiO_2$ interface. Subsequently, it produces a different capacitance value in depletion mode for a given voltage applied between its terminals. For a given dose of radiation, the capacitor-based detectors' sensitivity relies on the dielectric layer's thickness and the trapped charge density near the interface. Recently, the development of FDSOI technology has enabled the varactors to achieve high sensitivity and capacitance per unit area. In the presence of ionizing radiation, the thick buried-oxide (BOX) layer beneath the varactor elements collects the radiation-induced charges [83]. The resulting shift in capacitance value is measured using a readout circuit. Typically, an LC resonator equipped with the radiation-sensing capacitor provides a frequency output and the change of which is proportional to the accumulated dose [84]. FDSOI-based radiation-sensitive varactors are very effective in low-dose measurement with very high sensitivity [85].

2.3   Sensor Performance

Radiation sensors, namely, both dosimeters and particle detectors, are very effective tools to measure, evaluate, or characterize various radiation parameters (energy spectra, fluence, LET, etc.) and the effects (absorbed dose, bit-upsets, etc.). Semiconductor technologies have enabled sensor systems with the flexibility to be integrated with microelectronics and achieve better sensitivity with improved signal processing techniques. Semiconductor-based sensors are intrinsically reusable but suffer from a gradual loss of sensitivity caused by radiation-induced damages (crystal defects, leakage current, etc.) during their operation time. For a given radiation environment, the increasing leakage current can be compensated by using cooling, and the performance degradation may be restored by thermal annealing, thereby extending the sensor's lifespan. In the case of dosimetry, semiconductor-based sensors are generally used for relative measurement and need calibration with respect to absolute standards to ensure accuracy in readings.

A comparative study of the semiconductor-based radiation sensors based on their different features, methods, and technologies is provided in Table 2.1. Radiation-sensitive MOSFETs such as RADFETs and FOXFETs primarily employ DC measurement techniques, namely, V_{TH} measurement, and need simple readout architectures. However, static power consumption adds to the power overhead. The front-end electronics are very complex in the diode-based detectors operating in pulse or integration mode. FGMOS structures, which require charge injectors and varactors using AC methods, employ a similar level of complexity. As mentioned in Table 2.1, the most complex signal processing techniques and readout algorithms are employed in 3D NAND flash and resistive random-access memory (ReRAM)-based radiation sensors. Consequently, it increases power overhead, but not in the case of memristor-based ReRAM, as the memory array can be configured in passive mode with very low power consumption. Among the semiconductor-based sensors, the FGMOS devices, ReRAM, and FDSOI varactors exhibit the highest sensitivity toward radiation dose. Nevertheless, the measurand parameters exhibit non-linearity with respect to the accumulated dose, and the effective range of dose for which they could be used is significantly lower. These sensors are intended for low-dose applications like occupational dosimetry or therapeutic procedures. However, the FGMOS-based devices, as demonstrated in [77], support repeated measurements in auto-recharge mode and can be configured to record higher dose levels in accelerator environments. Although in such cases, longer readout times can accu-

Table 2.1 Comparison of semiconductor-based radiation sensors

Sensing element	Readout complexity	Power overhead	Dose sensitivity	Sense mode (SEE or TID)	CMOS technology
Silicon diode/photodiode	Medium	Low/medium	Medium	Both	Standard
RADFET	Low	Medium	Medium	TID	Custom
FOXFET	Low	Medium	Medium	TID	Custom
FGMOS	Medium	Low/medium	High	TID	Standard
SRAM	Medium	Medium	–	SEE	Standard
3D NAND flash	High	High	–	SEE	Custom 3D
MOS varactor	Medium	Low	Medium	TID	Standard
FDSOI varactor	Medium	Low	High	TID	FDSOI

mulate low-frequency flicker noise components, degrade the signal-to-noise ratio (SNR) performance, and reduce the radiation dosimeter's resolution.

2.4   Radiation Hardening and Mitigating Techniques

Researchers and radiation engineers try to opt for various radiation-hardening and mitigation techniques and strategies to protect the microelectronic devices and circuits against ionizing radiation to ensure reliable, robust operation. The primary way to achieve this is to use shielding with appropriate levels as applicable and affordable as possible. Although this technique is used extensively, particularly in the presence of high-dose radiation levels, it does not provide absolute immunity and mainly provides an abatement to radiation-induced damages [86]. Furthermore, in many space and terrestrial applications where the mass and size of payloads are limited, the shielding material required for safeguarding against high-energy radiation particles often adds to mission cost. It does not appear to be the optimal choice. Furthermore, the secondary radiation caused by the nuclear interactions between the primary radiation particles and the material inside may even increase the actual amount of energy deposited in the active regions of the component that is being shielded. Therefore, the preference for radiation engineers has always been to minimize shielding with various rad-hard technologies accompanied by circuit modification to achieve sufficient radiation hardness to sustain a safe operation. However, it is worth mentioning that the numerous radiation-hardening and mitigation methods do not imply immunity but rather robustness. These methods mainly weaken the radiation-induced damages or lower the degradation rate to extend the operation lifetime under specified limits.

Generally, two methods are primarily considered for radiation-hardening and mitigation purposes, which can be used separately or in combination to extend the radiation tolerance. The first method, known as radiation hardening by process (RHBP), focuses on modifying the semiconductor technology to reduce the radiation sensitivity of various physical properties. RHBP methods are advantageous because they can provide an intrinsic radiation tolerance without circuit modification, thereby minimizing designers' effort and development time. Although most RHBP-based solutions use existing mask sets with minor modifications, the added features and changes add to the cost and, therefore, appear to be the less preferred choice for commercial fabrication facilities. The second method is radiation hardening by design (RHBD) which looks for design-level mitigation techniques ranging from circuit-level modification to layout-level alterations to system-level redundancy algorithms. However, the drawback is the added design complexity with increased power consumption. Achieving sufficient radiation tolerance levels using RHBD techniques alone is often challenging while optimizing the system's performance in energy-constrained area-efficient applications. Therefore, the two methods are often combined to maximize the radiation tolerance while minimizing the development cost and designers' effort.

Various methods exist based on RHBP and RHBD techniques to mitigate radiation (TID and SEE)-induced damage or parameter variations. Typically, to achieve a system's or circuit's radiation hardness, a designer must first consider a suitable hardened technology in line with the specified requirements. Subsequently, further improvement can be attempted by combining it with various RHBD choices (layout level, circuit level, and system level). A more detailed overview of various radiation-hardening techniques (RHBP and RHBD) is provided in the following sections.

2.4.1   Radiation Hardening by Process (RHBP)

Process Hardening for TID

The extensive use of MOSFET devices for integrated circuit implementations has established TID as the dominant cumulative dose effect compared to DD in BJT devices. The radiation sensitivity of MOS devices' parameters depends on numerous factors involving the fabrication process, dielectric materials, device geometry, aging as well as operating conditions (electric field and temperature) and irradiation conditions (dose rate, dose, and ionizing source). Despite many complexities, the significant factors causing radiation-induced damages can be attributed to charge buildup in gate-oxide regions and Si-SiO_2 interfaces. In general, RHBP techniques for TID focuses on increasing electron trapping, reducing hole trapping, and increasing the quality of oxides. TID resilience by electron trapping can be achieved by implanting elements like Al, Si, P, Fl, and As into oxide layers.

As illustrated earlier, the radiation-induced shifts in physical parameters (threshold voltage, subthreshold slope, transconductance, noise, etc.) of MOSFET devices can be minimized by adopting thinner gate oxides. The radiation-induced charges in thinner (<10 nm) gate oxide (SiO_2) mostly get swept away due to tunneling. Bulk-CMOS with sub-100 nm channel lengths has achieved significant immunity against the gate-oxide-related radiation issue. Comparatively, oxide and interface traps in FOX region and excess leakage current through the device-isolation region are popularly identified as the radiation-induced short-channel effect (RISCE) and radiation-induced narrow-channel effect (RINCE) effects [87] which have appeared to be major contributing factors for TID-induced variations. To overcome these issues, there exist various rad-hard structural solutions like buried guard ring, heavily-doped-buried-layer (HDBL) under local-oxidation-of-silicon (LOCOS) isolation interface, and "corner rounding" etch process to harden the parasitic sidewall devices under shallow-trench isolation (STI) region. Parameter shifts caused by edge leakage through parasitic can also be reduced by introducing a parasitic isolation device ion implant around the active areas.

Process Hardening for SEEs

SEE mitigation can be achieved either by minimizing the amount of charge collected (Q_{coll}) at metallurgical junctions or increasing the critical charge (Q_{crit}) required to produce SEEs (SETs, SEUs, SELs, etc.). The reduction in Q_{coll} can be done by adopting one of the substrate-hardened technologies like silicon-on-insulator (SOI), silicon-on-sapphire (SOS), and silicon-on-diamond (SOD) among which SOI/SOS-based processes have been at the forefront for usage in space applications. There the substrate region is isolated from active devices using a BOX layer, which effectively reduces the e^--h^+ generation as well as charged carrier diffusion to active devices.

As the technology evolved, the successive reduction in feature size has also resulted in a significant decrease in Q_{coll}, which should have improved the SEE robustness. However, the reduction in Q_{crit} in line with voltage scaling and shrinking node capacitance has aggravated the problem. The increasing sensitivity of SEE was very prominent until the late 90s and reduced a bit as aggressive voltage scaling reached a limit in sub-100 nm nodes. However, in the last decade, the ever-increasing demand for higher functionalities and packed densities in CMOS ICs have had a negative impact on radiation-induced failures. SOI-based technologies can effectively reduce the occurrences of both SEUs and SELs. The STI regions reaching BOX eventually isolate the n- and p-wells in CMOS structures and inhibit parasitic p-n-p-n paths to prevent SELs. Another effective method to reduce SEL occurrence is to reduce substrate resistance using doping. The decrease in substrate resistance effectively drops carrier lifetimes, lowering the gain of parasitic feedback that can generate SEL. Usually, in such cases, an epitaxial layer is grown over the highly doped substrate to match baseline parameters. Besides substrate modifications, additional well structures can also be introduced to prevent SELs. Compared to dual well, triple-well structures are more robust against SELs. The deep N-well prevents the charge collection from radiation events that can lead to parasitic turn-ons. However, the effectiveness of SEL mitigation depends on well depths, operating voltages, and junction characteristics. A summary of popular and most prominent RHBP-based techniques is given in Table 2.2.

Table 2.2 Summary of RHBP mitigation techniques

Mitigation techniques	Targeted radiation effects			
	TID	SET	SEU	SEL
Implantation into oxides	X			
Buried guard rings	X			X
Buried layers	X		X	X
SOI/SOS		X	X	X
Epitaxial layers				X
Triple wells			X	X

2.4.2 Radiation Hardening by Design (RHBD)

Mitigation Techniques for TID

Over time, commercial CMOS technologies have evolved, and the resulting "TID radiation hardness by serendipity" [86] has appeared to be beneficial. However, the tolerance levels tend to vary with processes at similar technology nodes produced by different manufacturers and therefore need characterization data for consideration in the design phase. The TID hardness of advanced CMOS technologies mostly suffers from edge leakages and imperfections in the device-isolation regions. Therefore, RHBD by layout modifications, which leverages the physical and spatial properties of radiation effects, is considered one of the primary methods to improve TID robustness. As shown in Fig. 2.8a, the enclosed layout transistors (ELT) without any active diffusion edges separating source and drain can reduce the radiation-induced edge-leakage currents along the STI. For digital implementation, the drain in the inner region is preferred as it leads to lower drain-to-source capacitance. However, for use in analog circuits, the opposite is preferred as the output conductance reduces with reduced current density. Another alternative for eliminating edge-leakage effects is ring-source devices, as shown in Fig. 2.8b. Generally, ring source is preferred, as the ring-drain structure can significantly increase Miller capacitance. Besides, the area penalty of these modified layout devices lacks accurate modeling and therefore requires custom-cell libraries and spice models. Compared to ELTs, the area penalty for ring-source devices is less, so they can be used for designs with minimum-size devices. In the case of analog implementations requiring wider devices with higher driving strengths, the area penalty for ELTs is found to be less compared to standard designs. Wider devices ($>1 \, \mu$m) are comparatively robust in terms of TID-induced edge leakages and, therefore, preferred in place of ELTs which relieve the engineers from extra design effort without much compromising on radiation tolerance. An alternative way to mitigate the inter-device leakage is to encircle the active devices with a p+ diffusion ring.

Based on experimental data [87], NMOS devices are more resilient to radiation-induced damages than PMOS devices. In PMOS, the effects of charge trapped in bulk oxide and interface regions add up, whereas, in NMOS, those effects are opposite and, therefore, partially compensate for each other. In addition, the accumulated positive charges in the spacer oxide in the lightly-doped-drain (LDD) regions affect the PMOS and NMOS devices differently, which is identified as the radiation-induced short channel effect (RISCE). In the case of PMOS, the net effective doping in the LDD region tends to decrease, especially at the source side, causing the source resistance to increase. Thus, the effective transconductance of the device reduces. However, in the case of NMOS, the net effective doping in the LDD region tends to increase, which has a limited impact on the current flowing through the devices. Therefore, the designers could opt for NMOS-dominated architectures at the circuit level to minimize radiation-induced variations in circuit parameters. Under ionizing dose, the V_{TH} of PMOS decreases in magnitude, eventually pushing the device to complete cutoff, whereas the opposite happens with NMOS but at a slower pace. Designers could

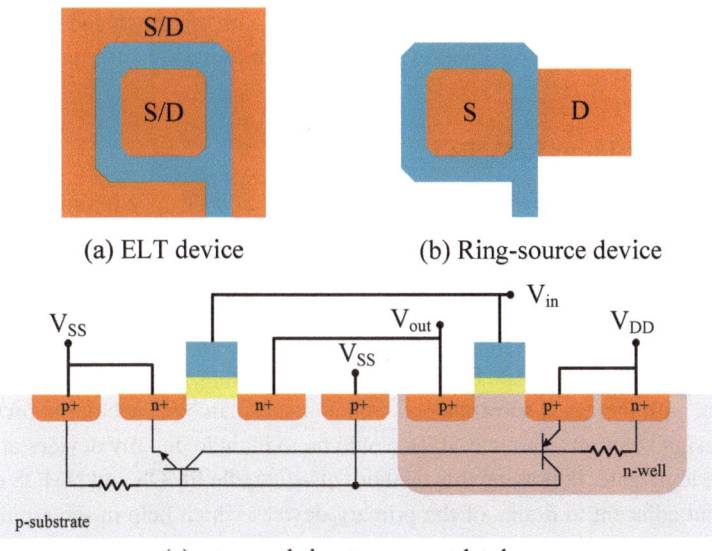

(a) ELT device   (b) Ring-source device

(c) p+ guard ring to prevent latch-ups

Fig. 2.8 Layout-level solutions for RHBD implementations

combine NMOS and PMOS devices with different flavors of V_{TH} parameters to elongate circuits' operation.

At the system level, the significant variations in performance parameters due to radiation can be accommodated by incorporating digitally assisted configuration blocks. Furthermore, the performance variation can be compensated using digitally assisted calibration. If the degradation rate and encountered radiation dose levels are high, one can opt for background calibration, but that includes added complexity and power overhead. If the system's requirements permit, an alternate design solution particularly applicable for low-dose levels could be foreground calibration which would interrupt the system and do the calibration only when it crosses the specified limits. In the case of analog circuits, the variation in DC parameters, the offset drift, and increasing 1/f noise in relation to the ionizing dose can be compensated using dynamic cancellation techniques like chopping, correlated double sampling, etc. Moreover, closed-loop architectures with negative feedback, like PLLs and DLLs, are relatively robust to TID and can sustain operation until a sufficient dose level is reached [17]. Unlike active devices, passive devices are less affected by radiation, so circuit specifications largely dependent on passive elements are quite robust under ionizing doses. For instance, a VCO's frequency undergoes minimal changes compared to that of a ring oscillator. Moreover, the replacement of the varactors in VCO with metal-oxide-metal (MOM) capacitors as in digitally controlled oscillators (DCOs) further improves the TID resilience. Analog circuit specifications such as negative-feedback amplifier's gain depending on ratios of passive devices or current-mirror's gain relying on device geometries are robust and radiation hardened. High-precision circuits with mismatch and variability-aware designs typically show high resilience against radiation-induced performance drift.

Mitigation Techniques for SEEs

As mentioned earlier, the probability of occurrence of SETs can be minimized by increasing Q_{crit} or reducing Q_{coll}. Typically, mixed-signal circuits use "brute force" methods to increase Q_{crit}. However, the area and power are often sacrificed for increased capacitance, wider devices, and increased drive currents. Several modifications like ELT and ring-source structures can be attempted at the layout level to reduce Q_{coll} collected at a device junction for minimizing SETs and SEUs. Further layout modification, such as guard rings as shown in Fig. 2.8c, with increased substrate and well contacts, can be beneficial to prevent SELs. Another design solution to reduce SETs would be to include dummy devices at the output node of the logic gate. Following this method, off-state idle PMOS and NMOS devices are put in layout adjacent to drains of the primary devices which help in the rapid collection of radiation-induced charges. In the case of analog implementations, the radiation-sensitive nodes can be interleaved with the ones with minimal effects while maintaining the packing density. Devices can be implemented with differential-charge-cancellation (DCC) layouts for differential circuits. Following this method, differential pair transistors are placed near such that the charge sharing can be maximized in the event of a radiation strike to improve SET performance.

At the circuit level, there exist several methods to reduce SETs. First, circuits can be implemented with lower bandwidths while reducing high-impedance nodes to prevent SET generation. Moreover, averaging analog voltages or currents at critical paths can be attempted to reduce the spurious transients in magnitude. Including a resistor in series with signal propagation or using low-pass and band-pass filters at critical nodes can effectively filter out high-frequency transients resulting from radiation events. However, one must be careful about the circuit's performance and area penalty. Differential switched-capacitor circuits vulnerable to transients can be hardened against SETs by splitting the input signal path into several parallel paths attempting redundancy. For digital circuits, several fault-tolerant techniques can be attempted at circuits' architecture levels to mitigate SEEs. As illustrated in Fig. 2.9a, a spatial redundancy method that uses a replica of radiation-sensitive blocks to detect discrepancies is usually very effective in masking SETs and SEUs. TMR-based spatial redundancy techniques employed locally or globally can detect and correct errors. Instead of the TMR circuits with thrice area penalty and power consumption, dual modular redundancy (DMR) can be tried with duplex architecture but is usually limited to error detection only. Spatial redundancy methods with an area overhead add to the power consumption and timing degradation due to increased parasitic contributions and interconnect delays. Moreover, as depicted in Fig. 2.9b, a temporal redundancy that samples the signals at different instants and uses them with voting circuitry can be included with spatial redundancy to improve resilience against SETs and SEUs further. In digital circuits, the propagation of SETs can be prevented using several methods, like logical masking and electrical and temporal masking. Another solution is the Muller C-element, as shown in Fig. 2.9c, with a differential delay between the data paths, which could remove any transients with a width less than the delay. Memory units such as SRAM cells, flipflops, and latches account for significant portions of the digital circuits, and they are vulnerable to radiation, particularly SEUs. For mitigation, designers have come up with several flavors of radiation-hardened storage cells like IBM hardened, heavy-ion tolerant (HIT), dual-interlocked storage-cell (DICE), and Whitaker's and Liu's SEU-hardened memory cell. Typically, these hardened memory units accomplish redundancy by combining the data storage with additional feedback paths to correct and restore the corrupted nodes. For instance, DICE-flipflops, as shown in Fig. 2.9d, consist of four units of inverters, and in the case of each inverter, the NMOS and PMOS devices are individually controlled by two adjoining units keeping the same logic state, and thereby achieving redundancy.

Considering today's ICs with packed densities and numerous design features, the scope of hardware redundancy is limited and not always affordable, considering the area penalty and power overhead. Therefore, for complex designs, particularly in the case of processors and SoCs, radiation hardening is attempted at the system level with the use of error-correcting codes (parity check, cyclic redundancy, Hamming codes, Reed-Solomon codes, etc.), watchdog timers as well as a set of techniques identified as software-implemented fault tolerance (SIFT). Compared to hardware-level redundancy, the main drawback is the timing overhead which affects the system's execution timeline. Typically, system-level mitigation

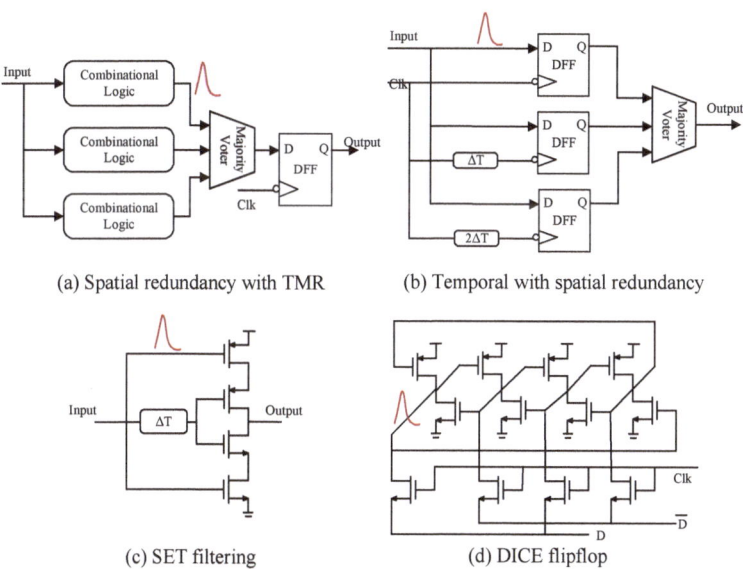

(a) Spatial redundancy with TMR   (b) Temporal with spatial redundancy

(c) SET filtering   (d) DICE flipflop

Fig. 2.9 Circuit-level solutions for RHBD implementations

Table 2.3 Summary of RHBD mitigation techniques

Mitigation techniques		Targeted radiation effects				
		TID	SET	SEU	SEL	MBU
Layout level	Edgeless/ELT	X				
	Well contacts & Guard rings	X			X	
	Dummy NMOS/PMOS	X		X	X	
	Transistor W/L ratio		X	X	X	
	Differential design/DCC layout		X	X		
	Node separation/interleaving		X	X		
Circuit level	Analog averaging		X	X		
	Resistive decoupling		X	X		
	Filtering/low bandwidth		X			
	Reducing high-impedance nodes		X			
	Dual/split path hardening		X			
	Rad-hard memory (HIT, DICE, Whitaker's, etc.)			X		
System level	Latch-up current monitoring		X		X	
	Error-correcting codes (Parity, Hamming, etc.)			X		
	Watchdog timers	X	X	X		
	Data scrubbing			X		X
	SIFT techniques (instruction/task)		X	X		X

techniques are still vulnerable to SETs in clock tree and reset path, and therefore more localized hardware-level mitigation can be attempted to further improvements. A summary of popular and most prominent RHBD-based design solutions is provided in Table 2.3.

2.5   Conclusion

This chapter briefly presents the fundamentals of ionizing radiation while mentioning the radiation mechanisms and sources. It has summarized the semiconductor-based radiation sensors (dosimeter and particle detectors) developed for radiation (TID and SEEs) measurement in the last few decades. Under radiation's impact, the electronic systems' performance degrades throughout the operation, and eventually the system fails in extreme conditions.

The radiation dosimeters help in qualifying the radiation hardness of the electronic systems, as well as monitoring and characterizing the radiation field for accurate modeling and implementing mitigation steps. Particle detectors primarily identify radiation-induced logic errors, memory upsets, and interrupts in electronic systems and subsequently help in error correction to ensure reliability and resilience against harsh radiation.

The chapter has also discussed various radiation-hardening and mitigation techniques for process changes, layout modifications, and circuit- and system-level implementations. Typically, TID resilience can be attempted by various RHBP techniques, primarily focusing on reducing hole trapping or increasing electron trapping and the quality of oxides. Further improvements can be targeted by RHBD techniques like adopting layout or design solutions that minimize edge leakages or self-compensate radiation-induced performance variations. Circuits primarily relying on passive elements also add to the TID immunity. In the case of SEEs, highly conductive substrates or buried layers help to minimize radiation-induced charge collection. In addition, active devices with sufficient well contacts and guard rings are helpful for latch-up prevention. Various redundancy techniques (spatial and temporal) are useful for digital circuits for SET and SEU mitigation. Furthermore, various radiation-hardened storage cells can further harden the circuits against radiation-induced failures. In the early days, RHBP-based methods were primarily targeted but always faced challenges due to affordability and the associated cost of process modifications. The scaling of CMOS technologies has helped to attain intrinsic radiation hardening with commercial processes. However, the situation turned adverse over time in the case of SEEs. Designers largely depend on RHBD techniques to minimize the SEEs, and in recent times various software-level mitigation solutions have been very effective.

Basics of Time-Based Signal Processing

3

Abstract

Over time CMOS technologies are optimized with shrinking feature sizes to enable the implementation of high-performance integrated circuits (ICs) with packed densities and numerous configurable features. Recent trends insist on fully integrating analog, mixed signal, and digital design on the same IC. Digital circuit designers have leveraged the benefits of technology trends with enhanced switching speed and increased transistor density. However, despite supply scaling in sub-100 nm CMOS technologies, there has not been any significant improvement in total power consumption, and the power density has almost remained unaffected. On the other hand, shrinking voltage headroom, increasing noise, and drop in intrinsic gain and dynamic range with low-feature sizes have not been very helpful for analog and mixed-signal designers. They mostly opt for circuits with large areas and high power dissipations to meet performance requirements. There has been an increased interest in finding alternatives to traditional voltage-based signal processing as it is mainly dependent on power-hungry area inefficient operational-transconductance amplifiers OTAs. A viable emerging alternative is time-based signal processing which has the potential to achieve analog-like circuit functionalities with highly digital-like circuits in scaled technologies [88]. The basic idea has been to modulate analog voltage- or current-based information with square-wave signals such that signal amplitudes are represented by transition edges. Although this technique existed earlier, there has been increased interest and effort to explore various time-based circuit topologies in the last decade. As illustrated in Fig. 3.1, time-based architectures exploited reduced feature sizes, unlike voltage-based circuits with faster transition times, which eventually translates to higher time resolution or, in other words, increased quantization capability. Typically, the time-based signal processing as in practice is accomplished by either time-based arithmetic units [89–91] using pulse width and time difference as signals or ring oscillators, using frequency outputs as signals. This chapter discusses the

© The Author(s), under exclusive license to Springer Nature Switzerland AG 2024   35
A. Karmakar et al., *Integrated Time-Based Signal Processing Circuits for Harsh Radiation Environments*, Synthesis Lectures on Engineering, Science, and Technology,
https://doi.org/10.1007/978-3-031-40620-1_3

fundamentals of time-based signal processing. It briefly explains the working principles of various building blocks, focusing on circuits using time-based arithmetic units.

3.1   Representation of Time-Based Signals

In time-based signal processing, the information or amplitude of the signal is represented by pulse widths, time periods, or time differences between two transition edges (rising or falling). Like digital signals, the time-based signals alternate between two voltage levels (high: V_{DD} and low: GND). The signals' waveforms are repetitive; therefore, it is only feasible to represent sampled information at regular time intervals in discretized formats. Real-world signals like sensor outputs are mostly voltage- or current-based continuous signals, which are transformed to discrete values once sampled with respect to regular time intervals. The discrete samples are digitized and converted to the digital domain using ADCs. Although the time-based signals resemble digital two-voltage-level waveforms, they are analog-type information with discrete values. Therefore, most of the discrete-time techniques applicable to voltage- or current-based circuits can be used to process the time-based information. Typically voltage-to-time converters (VTCs) and time-to-voltage converters (TVCs) are needed for the transformation of the signals from voltage to time based and vice versa. Afterward, like voltage-domain counterparts, time-based discrete values are digitized using time-to-digital converters (TDCs) suitable for digital signal processors. Digital-to-time converters (DTCs) are helpful for the opposite transformation of the signals if required by the applications (Fig. 3.1).

Figure 3.2 shows that the time-based signals can be represented as bipolar or unipolar information. In the case of unipolar, the signal is represented by the pulse width or time period of a single waveform. In contrast, in the case of bipolar, two waveforms are used, and the time difference between the transition edges represents the pseudo-differential values. Due to the resemblance to digital waveforms, the time-based signals inherit the digital noise margin

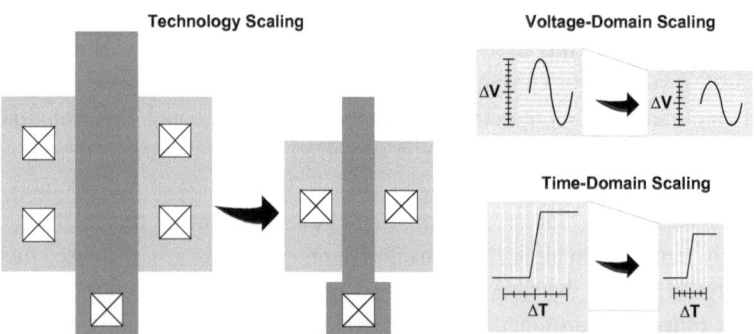

Fig. 3.1 Effects of technology scaling on signal processing

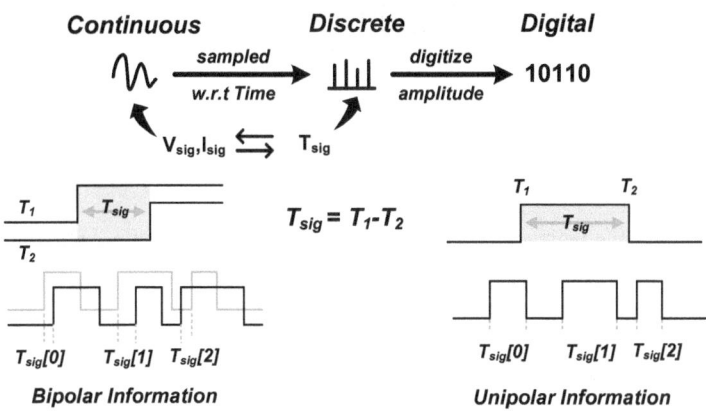

Signal Processing Flow

Fig. 3.2 Basics of time-domain signal processing

and are tolerant to noise when the voltage level stays at V_{DD} or GND. The precision of time-based information depends heavily on the accuracy of transition edges and is, therefore, very sensitive to noise contributions at and around transition edges. In theory, the supply limits the maximum amplitude of voltage signals, whereas the voltage noise determines the minimum limit. Therefore, the maximum achievable dynamic range reduces with shrinking voltage. On the contrary, the smallest amplitude of time-based signals is limited by the jitter contribution of the sampling clock. However, as long as one can measure, the maximum limit can be arbitrarily large (at least in theory). However, in practical clock-driven applications, this is mainly limited by the sampling clock period. The time difference or pulse width cannot be larger than the period within which it has been captured. While considering the delays produced by individual circuit blocks, the processing of the maximum amplitude of time-based signals is attempted by employing interleaved circuit topologies.

An overview of various building blocks used for time-based signal processing is given in the following sections.

3.2 Voltage-to-Time Converters

One of the basic building blocks of time-based signal processing is the VTC. As depicted in Fig. 3.3, VTCs use a real-world voltage signal as input and produce a time-variable $T_{sig}[n]$, which can be defined as the amount of time between a transition event occurring with respect to a reference transition time which in most cases can be attributed to the clock transitions. For single-ended circuits, a square waveform is produced where $T_{sig}[n]$ can be regarded as pulse widths proportional to $V_{in}(t)$. In the case of differential mode, $T_{sig}[n]$ is represented

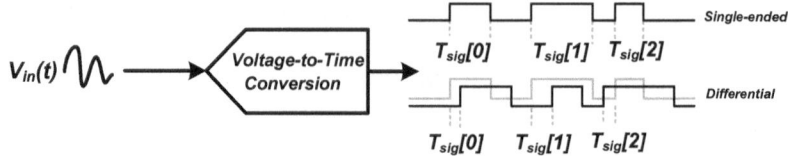

Fig. 3.3 Overview of voltage-to-time conversion

by the time difference between two transition events. The popular implementation types of VTCs are illustrated briefly in the following paragraphs.

3.2.1   Voltage-Controlled Delay-Based VTC

These types of VTCs include current-starved inverters (CSIs) and utilize inverter delay modulation techniques for voltage-to-time conversion. As illustrated in Fig. 3.4, the propagation delay of a CSI is modulated using current-controlling NMOS devices connected either in cascode [92] or in parallel [93]. As the clock goes low, the PMOS charges the output capacitance to V_{DD}, and the discharging happens as soon as the clock switches to high. In cascode configuration, the gate terminal of the current-starving NMOS transistor is driven by the sampled input voltage but with a limited range. The VTC-produced delays, essentially the output capacitor's discharging time, suffer from non-linear I_D/V_{GS} characteristics of the input-driven NMOS device. A resistor can be used with the cascoded NMOS for source degeneration to improve linearity, but that would limit the output swing. In a parallel configuration, an extra triode-mode NMOS is connected to a current-starving device to reduce non-linear distortion, which reduces the effective linear range at the output. Nevertheless, as depicted in Fig. 3.4, the linearity response of the delay produced in parallel configuration is better than that in cascode implementation.

3.2.2   Relaxation-Type VTC

The relaxation-type VTCs [94, 95] use charge domain techniques to convert the voltage to equivalent charge and then convert the stored charge to time-based variables like pulse width. At first, the VTCs as illustrated in Fig. 3.5 pre-charge the output capacitor to a reference voltage and then use a constant current source to charge or discharge it till it reaches the sampled input voltage level. The opposite way also exists, where the input voltage is first sampled on the capacitor and then charged or discharged until it reaches the reference voltage. For comparison, a differential-input analog comparator can be used, or a single logic gate can be utilized where the switching threshold voltage of the gate is used as a reference. Compared to earlier voltage-controlled delay-based VTCs, relaxation-type VTCs

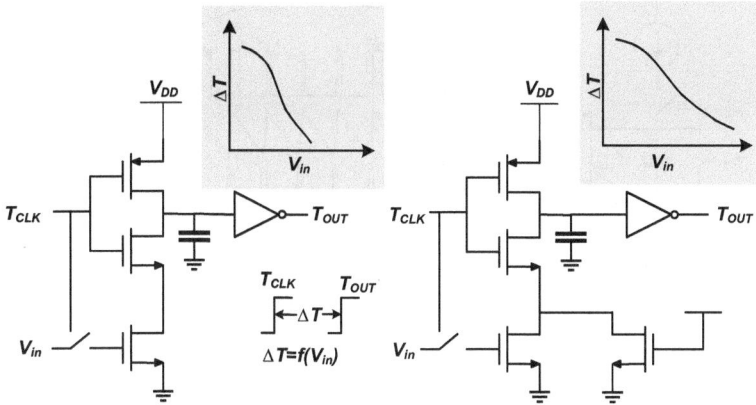

Fig. 3.4 Architectures of voltage-controlled-delay-based VTCs: (left) cascode configuration and (right) parallel configuration

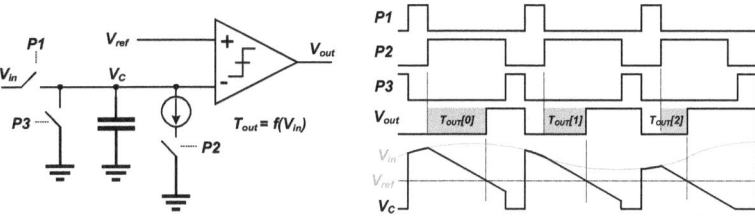

Fig. 3.5 Overview of the relaxation-type VTC

show better linearity as they do not suffer from non-linear voltage-to-current conversion like the former. The conversion gain of relaxation-type VTCs depends on the charging or discharging current. However, these VTCs use intensive analog circuits and therefore are not scalable like the earlier VTCs using CSIs.

3.2.3 Reference Voltage VTC

In this method, voltage-to-time conversion is done using the pulse-width modulation (PWM) technique [96–98]. As depicted in Fig. 3.6, a ramp voltage with a constant slope is generated at each clock cycle and compared with the sampled input voltage. Upon comparison, a pulse is produced at the output whose width is proportional to the input voltage. Like relaxation-type VTCs, these reference voltage VTCs provide good linearity and use analog circuit blocks like amplifiers and constant current generators. Furthermore, the analog comparator

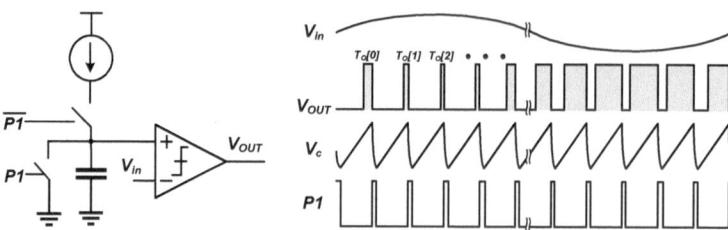

Fig. 3.6 Overview of the reference-type VTC

in the circuit provides a high input impedance and is attractive for many voltage-output sensor interfaces using time-based circuits.

The various types of VTCs described earlier can also be used to implement differential voltage-to-time conversion. The differential VTCs are typically implemented in pseudo-differential manner, and each half contains a VTC unit. Mismatches between the VTCs, if present, give rise to time-based offset, affecting the working of time-based circuits. Therefore offset cancellation techniques like chopping are required to compensate time-based offset values.

3.3   Time-to-Digital Converters

Time-to-digital converters (TDCs) are one of the critical blocks used in time-based signal processing. Similar to ADCs in the voltage domain, TDCs transform the time-based signals, like the time difference between two discrete transition events to digital codes (Fig. 3.7). They mainly use delays as unit time references for time-based comparison and quantize the time-based signal into digital values. TDCs, in general, exploit the shrinking feature sizes with sharper transition edges, which translates to reduced gate delays, i.e., increased time resolution. TDCs can be broadly categorized as sampling TDCs and noise-shaping TDCs. Sampling TDCs use open-loop architectures to digitize the time differences directly and are, therefore, useful for single-shot measurements. These TDCs provide a very short conversion time, but the in-band quantization noise limits the dynamic range. Noise-shaping TDCs with closed-loop architectures transform in-band quantization noise to frequencies outside the signal bandwidth. Due to oversampling, the noise-shaping TDCs use multiple consecutive

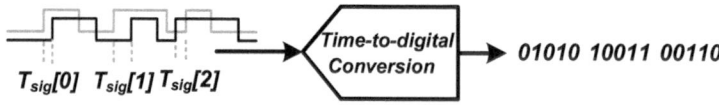

Fig. 3.7 Overview of time-to-digital conversion

samples and therefore do not suit single-shot measurements. A brief overview of popular, well-known architectures of TDCs is given in the following paragraphs.

3.3.1 Sampling TDC

Counter-Type TDC

A counter-type TDC converts a time-based signal T_{in} by counting the number of clock cycles within the duration of T_{in}. The resolution of the TDC equals the reference clock's time period; therefore, the counter is driven by a high-frequency low-jitter clock. As depicted in Fig. 3.8, the counter starts counting at the rising edge T_{start} and increments at each rising edge of the clock. It stops at the rising edge T_{stop} and latches the counter's value to the output. Counter-type TDCs provide a large dynamic range but at the cost of more power consumption and conversion time, which depends on the time duration of the input time-based signal.

Delay-Line/Flash TDC

Delay-line TDCs, also known as flash TDCs, work based on a principle similar to flash ADCs. As depicted in Fig. 3.9a, at the beginning of conversion in a flash TDC, a rising edge

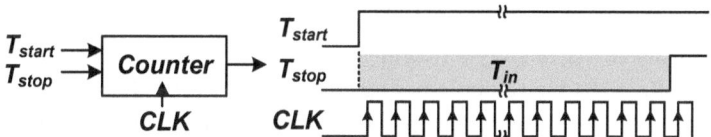

Fig. 3.8 Overview of a counter-type TDC

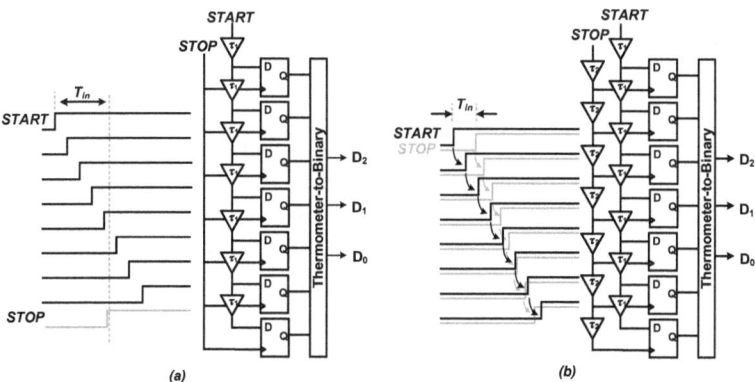

Fig. 3.9 Overview of the **a** flash and **b** vernier delay-line TDCs

is passed through a chain of delay elements which generates a number of time references in terms of delayed rising edges. When the "stop" transition event arrives, it is compared with respect to all the time references using D-flipflops. The D-flipflops produce logic high for all the rising edges, which leads to the "stop" event and logic low for the rest. The number of logic 1's at the output provides the number of delays which equals the time difference between the "start" and "stop" events. The outputs produced by the D-flipflops are thermometric and therefore converted to their binary equivalent. The resolution of the TDC is defined by the delays, which are produced mainly by simple digital buffers. The flash TDCs are very fast regarding time-to-digital conversion and therefore suit single-shot measurements. However, the TDCs have a limited dynamic range, and in a particular CMOS process, the resolution gets limited by the minimum gate delays. Scaled technologies with smaller feature sizes tend to help by reducing gate delays. To cover larger time intervals, the number of delay elements should be increased, eventually increasing area and power consumption.

Vernier Delay-Line TDC

As mentioned earlier, the quantization capability of flash TDC is limited by the smallest gate delays implemented in a particular CMOS process. Vernier delay-line TDCs are used to improve the resolution further with sub-gate delays. The working principle of the vernier delay-line TDC is similar to flash TDC; the only difference is that two sets of delay elements are used instead of one, as in the case of flash TDC. As depicted in Fig. 3.9b, both of the "start" and "stop" transition edges in the vernier delay-line TDC pass through two chains of delay elements having slightly different gate delays of τ_1 and τ_2 such that ($\tau_1 > \tau_2$). In that case, the effective time resolution of vernier delay-line TDC becomes $\Delta\tau = \tau_1 - \tau_2$. But the higher resolution is achieved only at the expense of area and power. Furthermore, unlike flash TDCs the outputs are not available immediately after the D-flipflops perform time-based comparisons. For a chain of N delay elements, the flipflop's output states are valid only after $N\tau_2$ time after the "stop" event arrives.

Pulse-Shrinking TDC

A pulse-shrinking TDC [99, 100] converts the pulse width to a digital value using a time-attenuator or a pulse-shrinking circuit. As depicted in Fig. 3.10, the input pulse width T_{in} propagates along the feedback loop, and each time it passes, the width of the pulse is reduced by a fixed amount ΔT. ΔT can be regarded as the resolution for time-digital conversion. At every clock cycle, a counter is incremented, and it continues until the pulse width reaches a minimum value and cannot trigger the counter. In the end, the output of the counter provides the digital value proportional to T_{in}. Similar to delay-line TDCs, the conversion time of pulse-shrinking TDCs depends on the input time difference.

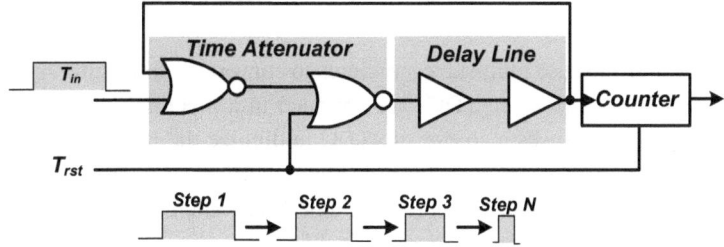

Fig. 3.10 Overview of pulse-shrinking TDCs

Successive Approximation TDC

Successive approximation register (SAR) TDCs use a binary-search algorithm [101, 102] to convert the input time-based signal to its digital counterpart. As illustrated in Fig. 3.11a, the successive approximation algorithm starts with the most significant bit (MSB) and compares the input pulse width T_{in} with 1/2 of the full-scale value T_{FS}. Based on whether $T_{in} > T_{FS}/2$ or not, the SAR logic sets the MSB bit to logic 1 or 0. Similar to SAR ADCs, if $T_{in} > T_{FS}/2$, T_{in} is compared to $T_{FS}/2 + T_{FS}/2^2$ or if $T_{in} < T_{FS}/2$, T_{in} is compared to $T_{FS}/2^2$. A time-based comparator usually makes the comparison and, based on the outcome, sets the corresponding bit to logic 1 or 0. For an N-bit SAR TDC, the time-based comparisons continue for N clock cycles to finish the quantization, and then the final digital value is latched to the output. Unlike voltage-domain SAR ADCs, the major challenge for implementing SAR TDC is to store a time-based variable and later retrieve it for comparison.

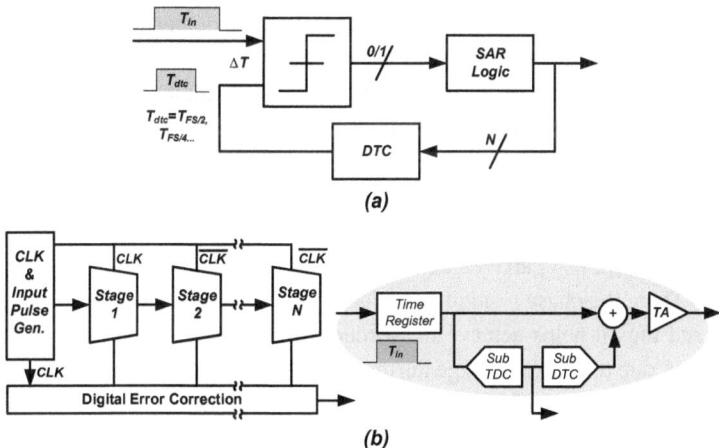

Fig. 3.11 Overview of successive approximation and pipeline TDC

Pipeline TDC

Pipeline TDCs [89, 103] use multistage structures to convert the time difference to a digital value. All the stages of pipeline TDCs are arranged in a pipeline structure. As depicted in Fig. 3.11b, each stage uses a coarse sub-TDC to digitize the input time difference and subsequently find the residue time error after digitization. The time residue is amplified (2x) by stretching the time difference using a time-based amplifier and transferred to the next stage for further digitization. Due to the multistage conversion in a pipeline structure, these TDCs can achieve a high speed and high resolution but are unsuited for applications with limited power constraints. The circuit complexity increases linearly with the number of bits used in the output. The linearity during the time-to-digital conversion depends on the accuracy of the gain of the time-based amplifiers and therefore needs digitally assisted calibration.

3.3.2 Noise-Shaping TDC

Similar to the $\Delta\Sigma$-ADC, the noise-shaping TDC uses noise-shaping and oversampling techniques to reduce the in-band quantization noise and improve the dynamic range (DR). In principle, it accumulates the quantization errors over time and pre-distorts the input signals so that a time average of the output digital sequence over several clock cycles provides an estimate of the input signal. In time-based signal processing, the noise-shaping TDC converts the time-based signals to the digital domain and the quantization error information is represented by residual pulse width or time difference. Alternatively, the phase residue can also be used as quantization errors in the case of time-based implementations, particularly the ones using ring oscillator (RO) as the core elements. As depicted in Fig. 3.12, there exist several methods for noise-shaping TDCs and the most notable ones are based on, namely, (a) a gated-ring-oscillator (GRO), (b) a switched-ring-oscillator (SRO), (c) higher order multistage-noise-shaping (MASH), and (d) standard single-loop $\Delta\Sigma$ architectures.

   As shown in Fig. 3.12a, a high-frequency GRO is enabled to run over the input time period $T_{in}[n]$. A counter counts the number of transitions at the output of the GRO. At the end of the time period $T_{in}[n]$, the GRO stops and the output node holds the phase and the voltage till the next time period $T_{in}[n + 1]$ arrives. A differentiator $(1 - z^{-1})$ at the output gives the final digital value. There the phase residue is carried over or "accumulated" from one clock cycle to another, and thus it helps achieve first-order noise shaping of the in-band quantization noise. However, due to charge leakage during the "hold" phase, the accumulator mostly acts as a leaky integrator, and the GRO-based TDC could not achieve optimum performance. As depicted in Fig. 3.12b, the SRO-based TDC solves the leakage issue by letting the RO run continuously but with two different frequencies f_H and f_L. The RO runs with frequency f_H for the time period of $T_{in}[n]$ and f_L for the rest of the time. GRO- and SRO-based noise-shaping TDCs use large oversampling ratios and provide a large dynamic range but suffer from higher order stability issues. Higher order noise shaping can be achieved using

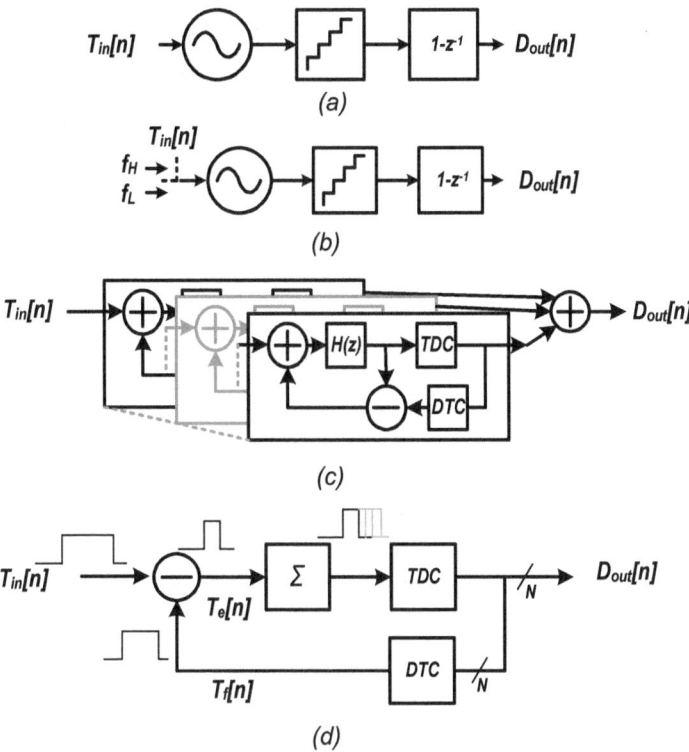

Fig. 3.12 Overview of architectures of different types of noise-shaping TDCs based on **a** a gated ring oscillator, **b** a switched ring oscillator, **c** MASH $\Delta\Sigma$, and **d** standard $\Delta\Sigma$

MASH-based architecture with GROs integrated within the loops. However, noise-shaping TDCs with MASH architectures are complex in design, dissipate more power, and are prone to spectral noise leakage from mismatch among cascaded branches. An alternate option is the standard single-loop $\Delta\Sigma$ modulation using time-based discrete-time filters as shown in Fig. 3.12d. The time-based arithmetic units accumulate residual pulse widths or time differences over the clock cycle. These $\Delta\Sigma$-TDCs use infinite-impulse response (IIR) or finite-impulse response (FIR)-based discrete-time filters to realize the noise-shaping filters.

3.4 Digital-to-Time Converters

The circuit maps a digital code to a time variable in a DTC. A digital-to-time operation is needed in applications such as time-based successive approximation TDCs and multi-bit noise-shaping TDCs. The DTC circuit assumes a similar role as that of a digital-to-analog converter in voltage-mode successive approximation ADCs or multi-bit $\Delta\Sigma$ modulators to

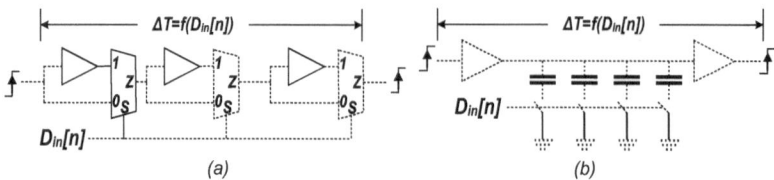

Fig. 3.13 Overview of architectures of different types of DTCs: **a** path-selection-based DTC and **b** variable-slope DTC

establish negative feedback and help in computing the quantization error during digitization. Recently, DTCs also find applications in the implementation of fractional-N ADPLLs [104] and multiplying DLL (MDLL) [105]. There are several architectures used to implement DTCs and the majority of them can be classified into three groups: (a) path-selection-based DTCs, (b) variable-slope DTCs, and (c) constant-slope DTCs. In the first group, as depicted in Fig. 3.13a, a chain of delay elements is constructed using simple digital buffers. The gate delay produced by the buffers represents the step size of digital-to-time conversion. Usually, the delays are controlled by a thermometric-encoded digital input. When disabled, the digital gates are bypassed as if they appear transparent and produce zero delays. In the second group of DTCs, a time delay with RC time constant is produced using an exponential charging or discharging profile [106]. To digitally control the RC delay, a capacitor array is enabled or disabled by the digital control word. A schematic representation of such a DTC is illustrated in Fig. 3.13b. These architectures are called "variable-slope" DTCs, considering the shape of the profile of charging or discharging. Unlike the path-selection-based DTCs, these DTCs can be easily implemented with binary-weighted devices (resistors or capacitors). Another major drawback of path-selection-based DTCs using delay chain is the high dynamic power.

There exists another architecture known as constant-slope DTC [107, 108] which uses a linear profile for charging or discharging capacitors. The constant-slope DTC can be implemented using relaxation-based VTC circuits. In this case, the sampled input voltage for the VTC circuit (see Fig. 3.5) is generated from a voltage-based DAC. Another option is to keep the input voltage constant and use a configurable current digital-to-analog converter (IDAC) to change the charging or discharging profile. Constant-slope DTCs act on minimizing the dependence of delays on input slew-rate variations and therefore showcase better linearity in comparison to the two DTCs mentioned above.

3.5 Challenges in Time-Based Implementations

Feature scaling with low-V_{DD} operation poses major challenges for analog voltage- or current-based circuits in terms of linearity, dynamic range, and noise levels. On the contrary, time-based circuits are energy-efficient OTA-less designs with increased time resolution despite having reduced voltage headroom. However, several challenges exist regarding the implementation of time-based circuits. The quantization capability of TDCs relies on minimum time delays produced by digital logic gates. Although smaller transistor sizes result in smaller gate delays, the random device mismatch increases with scaling. Therefore, devices with minimum sizes are not preferred for implementation which in turn deteriorates the effective time resolution. In time-based signal processing, the accuracy of information depends on time-based references like the delays produced by CSIs and clock frequency. The delays are subject to process variations and strongly relate to the supply voltage and temperature; any variations in their parts introduce errors in time-based signals and affect the time-based operation. Typically, DLLs and PLLs are utilized in time-based signal processing to generate accurate time-based references, but those add to the power, silicon area, and circuit complexity. Like voltage-domain signals, time-based signals are also susceptible to device-noise contribution at and around transition edges. Sharper transitions with a large slew rate reduce the gain for transforming voltage noise to time-based noise, but that comes at the expense of power. As devices become smaller, the performance of time-based circuits such as sampling TDCs for which quantization noise is the limiting factor tends to improve with reduced delays. For oversampling TDCs, the in-band time-based noise generated from devices and clock jitter is more dominant than quantization noise. Reduced area devices increase the noise contribution over the bandwidth and tend to decrease the maximum achievable dynamic range. Therefore, in such cases of time-based designs, the benefits from scaling are insignificant and the effective dynamic range of time-based circuits can be improved only at the expense of increased area and power. Another major challenge faced by time-based processing is the storage of time-based variables. Unlike voltage signals, a time-based variable cannot be stored in its original format. It is therefore converted to voltage and stored as charge in a capacitor. A time register stores the time-based signals, but their operation is limited in the presence of charge leakage due to capacitor non-idealities.

3.6 Conclusion

This chapter has briefly discussed the fundamentals of time-based signal processing while mentioning the working principle of various building blocks. Time-based signal processing circuits are primarily built using pseudo-digital logic gates, and the waveforms look like digital signals with two logic levels. However, unlike digital circuits, time-based circuits operate with discrete analog values encoded using pulse widths or time differences and can be represented as bipolar signals also. Unlike voltage mode, time-based circuits are energy-efficient OTA-less designs that exploit the shrinking feature sizes with increased time resolution despite having reduced voltage headroom. As discussed in this chapter, one of the basic building blocks is the VTC which converts the voltage-based signal into a time-based format based on the charging or discharging of a capacitor. Another well-known circuit is the TDC which digitizes the time-based signals using delay cells as reference. DTCs are used to generate digitally tunable pulse widths or time delays. The other circuits used in time-based signal processing are the time registers and amplifiers. These time-based arithmetic units are mainly used for time-based filter (IIR and FIR) implementations. One of the significant challenges for time-based implementation is the accuracy of delay cells and time references. Typically PLLs and DLLs are included in the time-based designs to generate low-jitter time references, but those add to circuit complexity and power. Time-based circuits exploit the transistor scaling with improved quantization capability and appear as significant competitors for voltage-based circuits during low-V_{DD} operation. However, the device-noise contribution also arises due to small feature sizes. Therefore in energy-constrained applications like sensor interfaces, challenges arise in time-based circuits to achieve small accurate time delays.

Radiation Assessment of Quadrature LC Oscillators

4

Abstract

This chapter presents a comprehensive assessment of the ionizing radiation-induced effects on the performance of quadrature phase LC-tank-based VCOs. Two different QVCOs capable of generating frequencies in the range of 2.5–2.9 GHz are implemented in a commercial 65 nm CMOS technology to target harsh radiation environments like space applications and HEP experiments. The architectures are based on the popular implementation of parallel-coupled and super-harmonic coupled QVCOs. The various performance metrics (oscillation frequency, quadrature phase, phase noise, frequency tuning range, and power consumption) of the two different QVCOs are evaluated with respect to a TID up to a level of approximately 100 Mrad (SiO_2) through X-ray irradiation. During irradiation, the electrical characterization of the prototype samples is performed under the biased condition at room temperature. The test setup of the TID experiment is discussed, and the results obtained are statistically analyzed in this chapter to perform a comparative study of the performance of the two different QVCOs and evaluate the effectiveness of the RHBD techniques employed in the implementations. The rest of the chapter is arranged as follows: Sect. 4.1 presents the prior art. Section 4.2 discusses the implementation of the QVCOs in detail. The test setup for X-ray irradiation is elaborated in Sect. 4.3 with a detailed description of the experimental results and a comparison of the performance study of the two different QVCOs. Section 4.4 concludes the chapter with a summary of the TID sensitivity study on the two QVCOs.

Parts of this chapter were adapted from the published open-access article in MDPI, Electronics journal [109].

4.1   Introduction

Various high-frequency designs, predominantly in radio-frequency- and millimeter-wave-based applications [110], require quadrature phase-shifted signals to enable wireless and wired communication systems. Wireless communication systems with integrated transceivers require quadrature signals for up- and down-conversion and eventually require accurate quadrature phase for effective image rejection in the baseband [111–113]. In the case of wired communication, QVCOs play a crucial role in multi-phase clock generation and, in particular, assist in the implementation of the half-rate clock and data recovery circuits [114, 115].

The use of QVCOs is extensive in various communication systems and half-rate CDR circuits. In modern days, these oscillators have also found indispensable use in high-speed communication in space (satellite communication, onboard space fiber network) applications and data transmission during HEP experiments. Considering the harsh radiation environment these oscillators are subjected to, the oscillators are required to sustain up to a TID level of several Mrad for space applications but several hundreds of Mrad for HEP experiments. The performance of LC-tank-based single VCOs operating under ionizing radiation is extensively studied in [15, 17]. The vulnerability of VCO architectures with respect to TID-induced effects for space applications and in HEP experiments is explored in [116].

In contrast, the effects of radiation-induced SEEs in VCOs are studied, and a few possible remedies are suggested with some design improvements in [16, 17, 117]. In comparison, the knowledge about the effects of ionizing radiation on the performance of QVCOs is limited due to scarcity in the number of studies [18, 19] done on QVCOs under radiation exposure. The SEE-induced effects are studied in [18] on QVCOs implemented in 65 nm bulk-CMOS technology. The TID-induced effects on a parallel-coupled QVCO implemented in 32 nm SOI CMOS technology are explored in [19]. The study reports on the degradation of several performance parameters (oscillation frequency, phase noise, and power consumption), but gives no insight into the radiation-induced effects on quadrature phase accuracy. This chapter focuses on giving more insights into the effects of radiation on QVCOs, with a particular interest in TID-induced variations.

4.2   Implementation

Numerous integrated design techniques [118, 119] for quadrature phase generation are reported in the literature to date and can be broadly identified as using (i) active RC oscillators [120], (ii) relaxation oscillators [121], (iii) ring oscillators [122, 123], (iv) poly-phase filters [124, 125], (v) frequency divide-by-2 circuits [126], (vi) cross-coupled QVCOs, and (vii) super-harmonic coupled QVCOs. Amid these, LC-tank oscillator-based options have prevailed as the design choice considering their superior performance in terms of phase noise and spectral purity within the given power budget. However, these are achieved with significant area penalties due to on-chip inductors. Although, in some low-frequency applications, limited by the area constraints and having not-so-stringent phase noise requirements, other non-LC-tank-based options are explored. LC-tank-based QVCOs typically contain two identical LC oscillators with the outputs cross coupled with each other. The quadrature outputs can be coupled through active devices [118, 127–129] or passive devices [130, 131]. The former method improves phase accuracy but at the cost of increased phase noise. In contrast, the latter showcases improved phase noise contribution but trade-off phase accuracy due to limited coupling strength. Conventionally, the coupling mechanism in the cross-coupled QVCOs follows either parallel [118, 129] or series [127] coupling schemes using the fundamental frequency component. The existing trade-off between phase noise and phase accuracy among these architectures can be eliminated by achieving quadrature phase locking using super-harmonic coupling, i.e., second harmonic injection at 180° out of phase at the oscillators' common-mode nodes [119].

For the TID experiment, two different topologies of QVCOs are implemented for quadrature phase generation: parallel-coupled QVCO (PQVCO) and super-harmonic coupled QVCO (SQVCO).

A schematic diagram of the PQVCO, as proposed in [132], with two identical core structures is shown in Fig. 4.1. Each oscillator core is implemented with complementary cross-coupled PMOS-NMOS transistor pairs (M_P, M_N) with an LC tank. The output of each oscillator is coupled to the other's output nodes parallelly using NMOS transistor (M_R) pairs. Based on the theory explained in [133], if the two oscillator cores are symmetrical and matched in terms of parasitics, the four outputs V_{A+}, V_{A-}, V_{B+}, V_{B-} produce quadrature phase-shifted differential I-phase and Q-phase (I and Q) output signals. The tunable capacitors in the LC-tank circuit are realized using n-well MOS varactors (C_{var}), and the inductor is implemented in top metal to maximize the inductor quality factor. As shown in Fig. 4.2a, the n-well MOS varactors are ac-coupled to the oscillator's output nodes using capacitors ($\sim 5 \times C_{var}$) and the gate voltage is varied using V_F connected through a resistor. This arrangement helps to reduce the sensitivity toward SETs [134]. Any charge collected

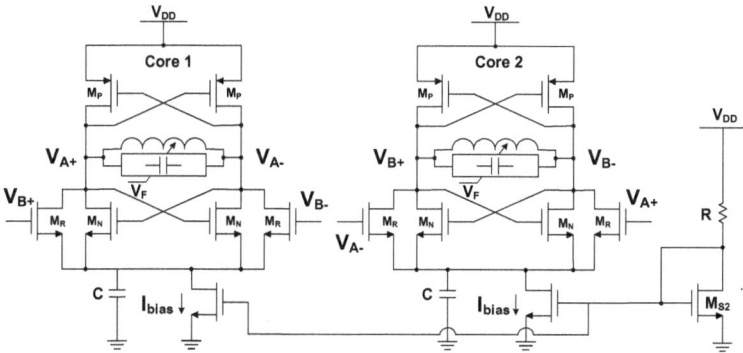

Fig. 4.1 A schematic representation of the parallel-coupled quadrature LC-tank VCO (PQVCO)

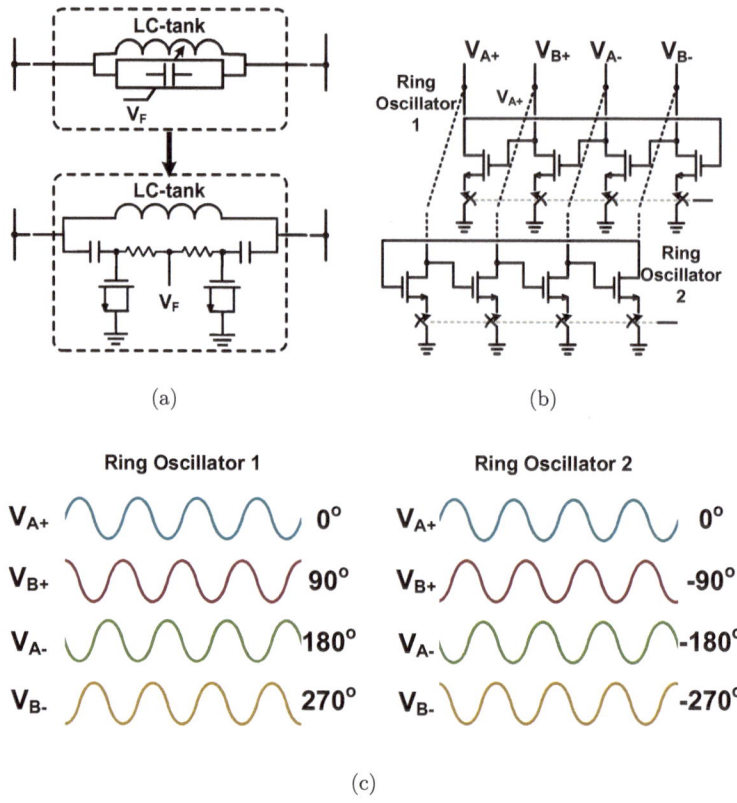

(a)   (b)

(c)

Fig. 4.2 **a** Tunable LC-tank circuit with ac-coupled NMOS varactor, **b** ring oscillators for phase lead/lag selection, and **c** oscillator generated output waveforms for different ring oscillator configurations

between the n-well and p-substrate interface finds a low-impedance path through the metal wires to the ground. However, this improvement pays a penalty of reduced frequency tuning range. Also, wider devices are chosen for core PMOS-NMOS pairs (M_P, M_N) to minimize radiation-induced performance variation [28], in particular, to reduce parasitic turn-on of the lateral devices due to STI-trapped charges. However, the use of wider devices in saturation results in increased power consumption. One could avoid wider devices with the use of ELT devices, but that required custom library or cells. The oscillators are designed to operate in full swing with a significant bias current (∼10 mA) at the edge of the current-limited region to achieve optimum phase noise performance within the power budget. The bias current is generated using typical current-mirror circuits formed with NMOS pairs. The bottom-biasing using NMOS is chosen instead of top-biased PMOS as the radiation-induced variation is less in NMOS devices compared to PMOS devices in 65 nm CMOS technology [28].

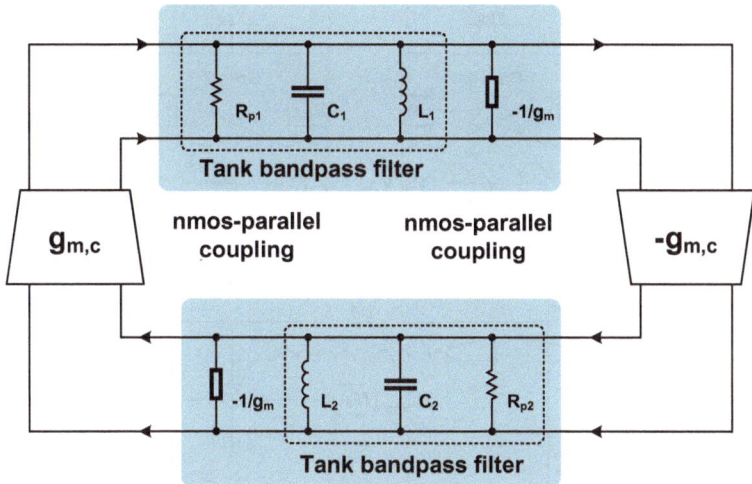

Fig. 4.3 A simplified block diagram of the PQVCO where two oscillator cores are coupled with each other through a transconductance ($g_{m,c}$) circuit

A simplified block diagram of the PQVCO cross coupled with each other through transconductance ($g_{m,c}$) formed by the coupling NMOS transistors (M_R) is shown in Fig. 4.3. The LC-tank circuits (R_p, C, L) form the tank band-pass filter at the resonance frequency and complementary PMOS-NMOS pairs (M_P, M_N) create the $-1/g_m$ cell to compensate for tank losses. Quadrature phase accuracy can be improved with stronger coupling strength $g_{m,c}$, but it leads to increased phase noise around the oscillation frequency. Therefore, the sizes of M_R are chosen smaller, which is 1/4 times of the core MOS devices (M_N) to minimize the phase noise contribution while maintaining the quadrature phase error within acceptable limits ($<1°$). Due to symmetry in operation in the identical core structures, the PQVCO will likely exhibit bi-modal oscillation [135]. However, in the presence of a mismatch in the tank circuits or coupling devices, one of the modes prevails and leads to oscillation with either 90° or −90° phase-shifted outputs. The ambiguity in phase is resolved by introducing ring oscillator structures [111, 136] at the output nodes. As shown in Fig. 4.2b, two ring oscillators are provided, and either one can direct the oscillation in either leading or lagging phases. The resulting waveforms of V_{A+}, V_{A-}, V_{B+}, V_{B-} based on different ring oscillator configurations are shown in Fig. 4.2c.

Figure 4.4 shows a schematic of the SQVCO implemented for the TID experiment based on the architecture proposed in [136]. As elaborated in Fig. 4.4, the two oscillator cores are identical, and the common-mode nodes are coupled at 180° out of phase through a center-tap inductor. Ideally, in a differential configuration, the odd harmonics of the oscillator circulate through the switching core MOS devices, and higher order even harmonics appear at the common-mode node of the oscillator core. As per the super-harmonic injection locking mechanism [136], each oscillator's second harmonics are mutually coupled at 180° out of

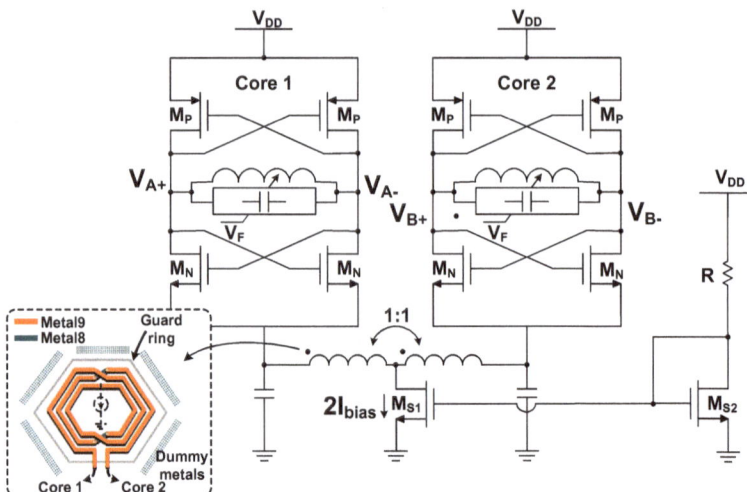

Fig. 4.4 A schematic representation of the super-harmonic coupled quadrature LC-tank VCO (SQVCO)

phase, and this leads to quadrature phase differences at the fundamental frequency component. In this architecture, a center-tap inductor is used in series with the bias circuit. Each half is configured to resonate with the parasitic capacitors at twice the fundamental frequency. The second harmonic at the common-mode nodes is coupled magnetically through each half of the inductor and maintains a 180° phase difference between each other. A layout-level representation of the center-tap inductor working as the coupling transformer is shown in the bottom-left corner of Fig. 4.4. The inductor is implemented in top metal with dummy metal fillings and guard rings to prevent substrate coupling.

Unlike the PQVCOs, the absence of coupling through active devices in this architecture breaks the trade-off between quadrature accuracy and phase noise contribution. Even though the sizes of the core MOS devices are identical in the PQVCO and SQVCO and both are driven by equal bias currents, the phase noise performance is better in the SQVCO as compared to PQVCO. The coupling inductor acts as a tail-noise filter [137]. It generates a high impedance at twice the fundamental frequency at resonance, reducing flicker noise up-conversion and thermal noise down-conversion around the oscillation frequency.

For better insights into the operation, a simplified block diagram of the SQVCO circuit components [111] is shown in Fig. 4.5. The LC-tank components (R_p, C, L) act as a band-pass filter centered at resonance frequency where the tank losses are compensated by the $-1/g_m$ cell created by the complementary PMOS-NMOS pairs (M_P, M_N). The common-mode node at the oscillator core acts as a frequency doubler and injects the signal to the other with an added phase shift of 180°. The injected out-of-phase second harmonic produces a quadrature phase shift while mixing with the fundamental frequency component. In this architecture, the quadrature accuracy largely depends on the tail node impedance

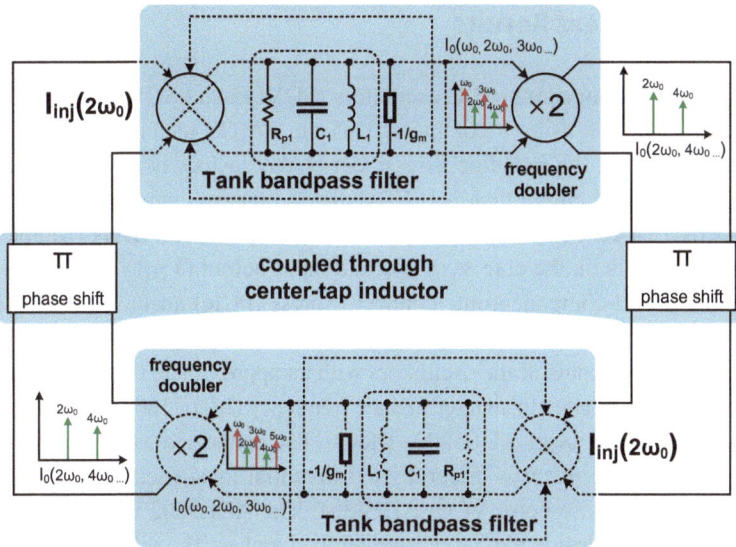

Fig. 4.5 A simplified block diagram of the SQVCO where two oscillator cores are coupled with each other through a center-tap inductor

and resonating signal strength at the tail inductor [136]. Similar to the PQVCO, two ring oscillators are provided at the output nodes, as shown in Fig. 4.2b, to resolve the quadrature phase ambiguity (lead or lag). In addition, the core PMOS-NMOS switching pairs and the bias circuit of both types of QVCOs, PQVCO and SQVCO, are implemented with p-type and n-type guard rings to prevent SEL-induced failures.

The off-chip driving strength of the differential output signal pairs from the oscillators is improved using cascaded CMOS digital buffers with increased driving strength. As shown in Fig. 4.6, two parallel chains of digital buffers are used in differential configuration with cross-coupled latches connected between the nodes for phase alignment.

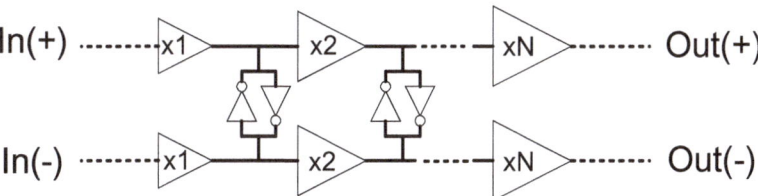

Fig. 4.6 Cascaded CMOS digital buffers used at the oscillator outputs

4.3   Measurement Results

The prototypes of the quadrature LC oscillators (PQVCO and SQVCO) are implemented together in a single die using a commercial 65 nm CMOS technology. An overview of the test setup used for the radiation assessment of the devices is provided in Fig. 4.7. It consists of two printed circuit boards (PCBs) and the measurement equipment controlled by a Raspberry Pi 4-based controller board. The device-under-test (DUT) samples with the decoupling capacitors on the core $V_{DD,core}$ and input-output ($V_{DD,IO}$) supplies are wire-bonded to peripheral-component-interconnect-express (PCIe) adapter boards. The micro-photograph of the fabricated die sample is shown in the top-right corner of Fig. 4.7. The buffered quadrature outputs of the oscillators with an approximate frequency range of 2.5–2.9 GHz are down-converted to an intermediate frequency (IF) of 500 MHz using an off-chip double-balanced active mixer ADL5802. The DUT is connected to the mixer through the PCIe interface, and the mixer is driven by a differential local oscillator signal (2.4 GHz) generated from low-phase noise VCO, ADF4351. The ADL5802 interface PCB with the DUT sample is placed inside the X-ray irradiation chamber. The differential IF quadrature output pairs from the onboard 1:1 balun transformer are brought out of the X-ray chamber to the Keysight DSA91304A oscilloscope using 2 m long sub-miniature version A (SMA) cables for frequency and quadrature phase measurement. The ADL5802 interface PCB and

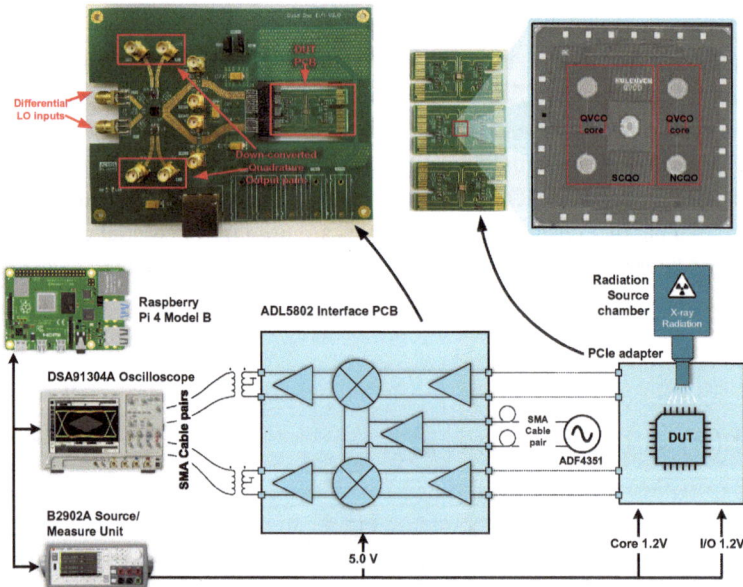

Fig. 4.7 Overview of test setup for radiation assessment of QVCOs and the micrograph of the die (2×2 mm^2) in the top-right corner

$V_{DD,IO}$ of the DUT are powered from fixed 5 V and 1.2 V supplies, respectively. One of the channels of the Keysight B2902A precision source/measure unit is used to provide 1.2 V to the $V_{DD,core}$ and measure the current consumed. At the same time, the other channel is used to tune the varactor input voltage during the measurement. Additionally, a PN9000 phase noise analyzer is utilized to measure the output phase noise of the free-running quadrature oscillator.

The radiation assessment of the DUT is conducted under biased conditions at room temperature. While testing, the DUT is placed at the center of the incident X-ray beam having a diameter of approximately 3 cm. The X-ray beam is generated from a 40 keV, 40 mA W-tube from Seifert, resulting in a dose rate of 36.78 krad/min. Before irradiation, the dose rate of the X-ray beam is calibrated using a PIN diode-based dose sensor. During irradiation, the various performance metrics (frequency, quadrature phase, core power) of the DUT are measured repeatedly up to a TID level of 100 Mrad (SiO_2).

The percentage variations of the measured output frequencies (f_{max}, f_{center}, and f_{min}) of PQVCO and SQVCO are shown in Fig. 4.8a, b, respectively. Before irradiation, the frequency of the PQVCO ranges from 2.524 to 2.786 GHz with a tuning range of 9.86%, and the frequency of the SQVCO ranges from 2.635 to 2.908 GHz with a tuning range of 9.88%. Similar to the TID experiments [17, 116], the oscillation frequency of both oscillators gradually increases with respect to TID, and an increment of approximately 0.5% can be observed at the center frequency. The increase in the frequency can be accounted for primarily due to the decrease in the transconductances (g_{mP}, g_{mN}) of the cross-coupled PMOS-NMOS pair with respect to TID [28]. As elaborated in [138], the frequency of oscillation can be expressed as

$$f_{OSC}^2 \approx f_0^2 \left(1 - \frac{1}{Q_L^2} + \frac{1}{Q_C^2} + \frac{C_{CM}}{C_{tank}}\left(1 - \frac{1}{g_m^2 \left|Z_{s,2}\right|^2}\right)\right), \qquad (4.1)$$

where f_o is the tank resonance frequency, Q_L is the inductor quality(Q) factor, Q_C is the capacitor Q factor, C_{CM} is the capacitance at the common-mode biasing node, C_{tank} is the tank capacitance, g_m is the effective transconductance of the cross-coupled pair, and $Z_{s,2}$ is the effective impedance at the common-mode biasing node at the second harmonic of the oscillation frequency.

In the oscillators (PQVCO, SQVCO), the NMOS varactors are ac-coupled and therefore are not affected by the level of output common-mode voltages. As the varactor voltage is tuned from 0 to V_{DD}, the varactors' gate voltage is always greater than the source and drain voltage, so they operate entirely in accumulation mode. The effects of oxide-trap and interface-trap charges on the varactors with respect to the bias voltage are analyzed in detail in [139]. During irradiation, the radiation-induced negative charges (oxide trap and interface trap) reduce the width of the depletion region, which in turn increases the capacitance of the varactors [116, 140]. This radiation-induced effect is counteracting the previous g_m induced

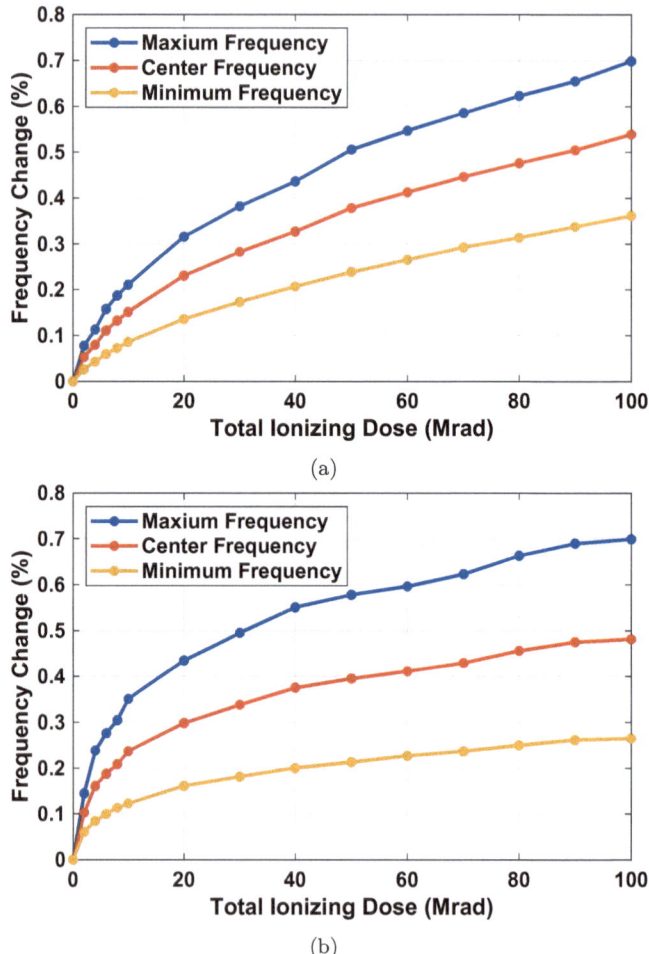

Fig. 4.8 Percentage frequency variations of **a** PQVCO and **b** SQVCO measured with respect to the TID of the order of 100 Mrad (SiO$_2$)

increase in frequency. However, as elaborated in Eq. 4.1, the combined effect of the increase in C_{tank} and the decrease in g_m results in an increase in the frequency of oscillation. In the case of f_{max}, the contribution of C_{tank} is less compared to that of f_{min}. For lower values of C_{tank}, the effect of g_m is much more dominant, and therefore the relative variation of f_{max} is more than f_{min}. In the case of the SQVCO, the effective impedance $Z_{s,2}$ at the common-mode biasing node is larger than the PQVCO because of the resonance of the coupling inductor. Therefore, the increase in the term $1/(g_m^2 |Z_{s,2}|^2)$ is less in the SQVCO than in the PQVCO, which results in a slightly larger relative variation of f_{min} in the PQVCO compared to the SQVCO. The variations of the tuning range of the oscillators, which can be expressed

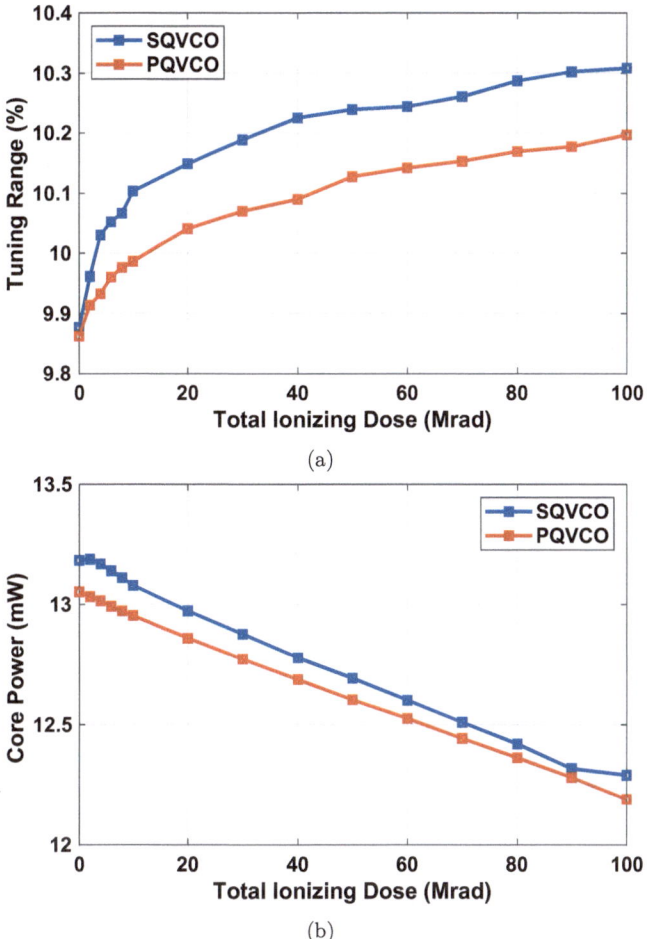

Fig. 4.9 a Tuning ranges, **b** core power consumption of the SQVCO and the PQVCO measured with respect to the TID of the order of 100 Mrad (SiO_2)

as $T_R = 100 \times (f_{max} - f_{min})/f_{center}$, are measured with respect to TID and shown in Fig. 4.9a. After 100 Mrad (SiO_2) TID, the relative variation of the tuning range of the PQVCO is 3.4% and that of the SQVCO is 4.4%. Despite having similar relative variations of f_{max} for both the oscillators, the relative variation of the tuning range of the PQVCO is less than the SQVCO primarily due to larger variations of f_{min} in the PQVCO compared to the SQVCO.

The variations of the dissipated core power of the two oscillators (SQVCO and PQVCO) are shown in Fig. 4.9b. Before irradiation, the core of the PQVCO consumes a power of 13 mW, whereas the core of the SQVCO consumes a power of 13.2 mW. The dissipated core power reduces with respect to TID, and after 100 Mrad (SiO_2), the core power changes around 7% for both oscillators. NMOS pairs in current-mirror formation with resistive pull-up are

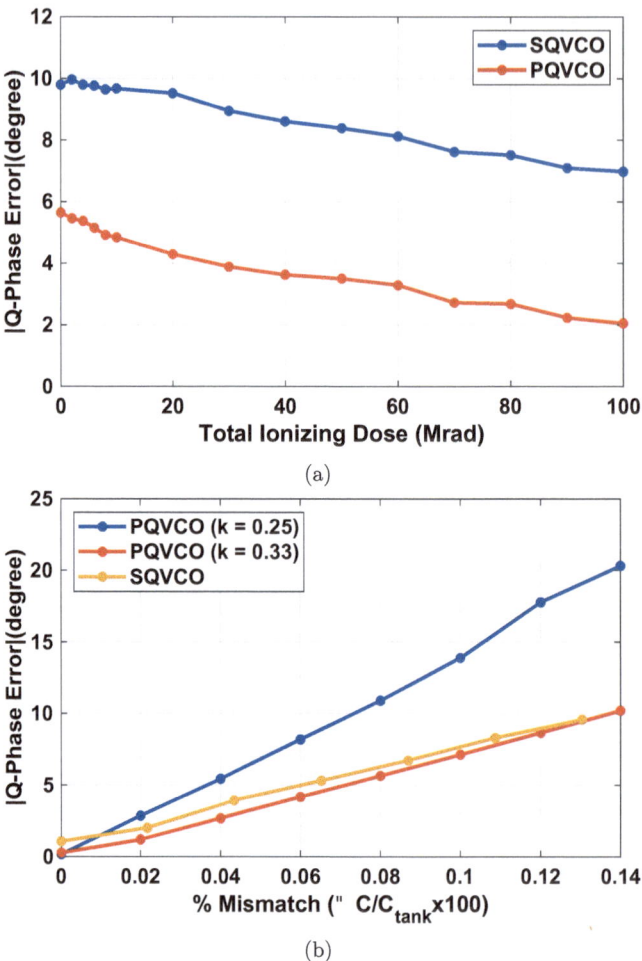

Fig. 4.10 a Variations of Q-phase error measured with respect to TID of 100 Mrad (SiO$_2$),
b simulated Q-phase error for the SQVCO and PQVCO with respect to different relative mismatch
in C_{tank}

used to provide the bias currents to the core of the oscillators. While exposed to radiation, the threshold voltage of the NMOS devices gradually decreases [28], which in turn results in an increased overdrive voltage and a decrease in the current through the resistive pull-up path. Due to the mirroring action, the core bias currents also undergo similar reductions. The reduction in bias current also corroborates an increment in oscillation frequency with respect to radiation. As elaborated in [138], the relationship between the oscillation frequency and the bias current is such that when the oscillator is biased at the edge of the current-limited region using a large bias current, a reduction in bias current shows a slight increase in frequency.

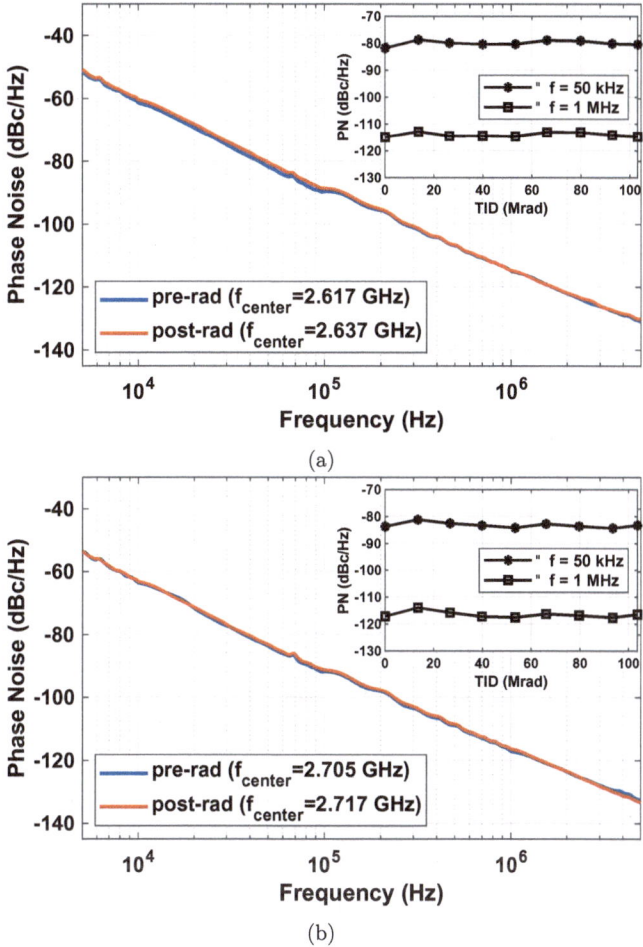

Fig. 4.11 Phase noise measured at f_{center} before (pre-rad) and after (post-rad) radiation exposure of the order of 100 Mrad (SiO$_2$) from the outputs of the **a** PQVCO and **b** SQVCO

The variations of the absolute Q-phase error of the two oscillators are measured with respect to radiation up to a level of 100 Mrad (SiO$_2$). Before irradiation, the Q-phase error is larger (10°) in the SQVCO compared to the Q-phase error (5.9°) of the PQVCO. The Q-phase error largely depends on the mismatch between the tank resonance of the mutually coupled oscillator cores and the coupling strength of the I/Q-phases. The variation of absolute Q-phase error of the two oscillators with respect to the mismatch of C_{tank} based on post-layout simulation results is shown in Fig. 4.10b. The Q-phase error decreases with the increasing mismatch between C_{tank} of the oscillator cores. In the case of the PQVCO, the change in phase error is much more rapid with respect to the mismatch for lower values of the coupling factor (K). During the radiation experiment, the capacitance of the NMOS varactor proportional to the tuning voltage increases with respect to the radiation dose [116, 140]. The increasing capacitances eventually reduce the relative mismatch between the C_{tank} capacitors of the oscillator cores. The effect can be seen under radiation as the Q-phase error tends to reduce for both the oscillators as depicted in Fig. 4.10b. As found from the simulation (Fig. 4.10b), increasing the coupling strength (K) diminishes the effects of mismatch.

The phase noise characteristics of the PQVCO and the SQVCO measured at frequency f_{center} are shown in Fig. 4.11a, b, respectively. The blue lines show the phase noise contributions measured prior to radiation exposure, and the red ones show the phase noise observed after 100 Mrad (SiO_2). As can be observed in Fig. 4.11a, b, the phase noise characteristics in the $1/f^2$ region (>300 kHz offset) do not undergo many variations. This is counter-intuitive, with a 7% reduction in the bias current. Here, the oscillators are biased at the edge of the current-limited region to produce full-swing outputs and achieve optimum phase noise performance. In this region of operation, the changing bias current has minimal effect on the oscillators' phase noise characteristics [141]. Although the thermal noise contribution of the cross-coupled core devices is supposed to increase with respect to TID [142], the overall phase noise characteristics in the $1/f^2$ region do not change much. This is primarily due to the wider device sizes used in the cross-coupled pairs, which reduce the thermal noise contribution in the phase noise characteristics. However, close observation can reveal that the close-in phase noise in the $1/f^3$ region (<300 kHz offset) increases slightly in the case of the PQVCO compared to the SQVCO. This is due to the extra pair of NMOS devices used for coupling the I- and Q-phases. The sizes of the devices are less compared to the core devices (coupling factor K $= 0.25$) and therefore contribute more 1/f noise. Overall, the phase noise characteristic of the SQVCO is slightly better than that of the PQVCO due to the tail-noise filter [137] in the form of the coupling inductor present in the SQVCO.

4.3.1 Performance Comparison with State-of-the-Art Designs

Performance comparison of the implemented QVCOs with respect to previously published VCOs and QVCOs, which are characterized by radiation exposure (TID), is presented in Table 4.1. Similar to [17], the implemented QVCOs in this work are targeted to operate in the s-band. The implemented QVCOs consume double power in comparison with [17] as two oscillator cores are coupled together to generate quadrature phases. Based on open-loop phase noise characteristics, the implemented QVCOs achieve more than 3 dB better phase noise at 1 MHz offset. However, [17] has a similar rate of variations of the frequency with respect to TID level compared to the implemented QVCOs. The oscillator in [144] has combined small varactors with large digitally switchable capacitor banks (64 units) and, therefore, could achieve better phase noise performance than the other reported designs. The

Table 4.1 Comparison with state-of-the-art TID evaluated VCOs and QVCOs

Reference	TNS'17 [17]	TNS'18 [134]	TNS'18 [143]	TCASI'19 [144]	TNS'21 [116]	TNS'17 [19]	This work	
Technology (nm)	65	65	65	65	65	32	65	
Type	VCO	VCO	VCO	VCO	VCO	QVCO	QVCO†	QVCO‡
Oscillator area (mm^2)	–	–	0.124	–	0.061	0.0484	0.458	0.367
Frequency (GHz)	2.2–3.2	2.5–2.65	4.8–6.0	4.9–5.2	5.4–6.8	20.1–20.7	2.6–2.9	2.5–2.8
Tuning range (%)	30**	5.8	4	5.9**	23**	3	9.9	
Phase Noise @1 MHz (dBc/Hz)§§	−110	−118	–	−122	−100	−99	−119	−115
VCO gain (MHz/V)	240**	–	1850	100	225	610	273	262
Power (mW)	6	1.8	18	34	2.85	12.8	13.2	13
FoM§ (dBc/Hz)	−171	−188.7	–	−180	−171.4	−176	−176.4	−172.2
Frequency change (%)	3.5	–	–	3.2	2.54	1.4	0.7	
TID tolerance (Mrad)	600	–	250	350	1000	0.5	100	

§ FoM = Phase Noise@$\Delta f - 20\log(\frac{f_o}{\Delta f}) + 10\log(\frac{Power}{1mW})$

† SQVCO ‡ PQVCO §§ Open-loop phase noise

** Varactors and digitally controlled capacitor banks used

QVCOs designed in this work have compromised the tuning range due to the AC-coupling of varactors but still achieve better tuning range compared to other implementations [19, 134, 143, 144]. The tuning ranges of [17, 116] are extended using digitally controlled capacitor banks. However, a trade-off exists between the tuning range and the phase noise characteristics, and therefore [17, 116] reports higher phase noise contributions. Prior to this work, as published in the literature to date, the design in [19] has been the sole instance of a TID study performed on QVCOs. Therefore, the results are included in the comparison, even though it is implemented using a 32 nm CMOS SOI technology different from the one used in this work. Although TID-induced variations tend to reduce with smaller feature sizes, the QVCOs reported in [19] show more frequency variation and less TID tolerance compared to the QVCOs implemented in this work.

4.4 Conclusion

The performance study and evaluation of radiation-induced effects on the prototypes of the QVCOs, PQVCO and SQVCO, are presented in this chapter. It shall help to understand the implementation's vulnerabilities and improve the design for future implementations. The samples of the prototypes are tested under X-ray radiation up to a level of 100 Mrad (SiO$_2$). The overall variations of the frequencies of the oscillators are less than 1%, and the change in tuning range is less than 5% after the radiation exposure, which makes them suitable to be integrated inside PLLs and half-rate CDRs. Although the bias current reduces by 7%, it has a limited contribution to the phase noise performance of the QVCOs. The most vulnerable performance metric is the Q-phase error of the oscillators. In this experiment, the decreasing relative mismatch between the C_{tank} improves the Q-phase error. However, for applications involving much higher frequencies, the value of C_{tank} used would be less, which may worsen the radiation-induced variations in the Q-phase. The solution would be to increase the coupling strength in the PQVCO to minimize the mismatch-induced variations at the expense of phase noise performance. In the case of SQVCO, extra effort is needed to produce matching layouts of the oscillator core.

Radiation-Hardened Delta-Sigma CDC

5

Abstract

This project implements a novel time-based capacitance-to-digital converter (CDC) fabricated in a commercial 65 nm CMOS technology. The radiation-hardened CDC is targeted for use with capacitive sensor interfaces in harsh radiation environments. It is expected to operate with minimal performance variation with respect to TID (up to 10 MGy-Si) and SEL tolerance up to 65 MeV.cm^2/mg. In particular, during HEP experiments, the relative humidity is measured to ensure the "dryness" of the Si-detectors before being cooled down below zero temperatures to avoid vapor condensation. Miniaturized integrated capacitive humidity sensors are installed along with the detector assembly to monitor the humidity readings and the dew point temperature. The CDC prototypes are fully designed using time-based digital-gate-like circuits. The CDC uses 1st order $\Delta\Sigma$ quantization noise shaping to improve the measured capacitance resolution. The prototype supports capacitance measurement within the range of 0–3.75 pF. The CDC samples achieve an effective-number-of-bit (ENOB) of 12.9 bits with an energy efficiency of 0.18 pJ/-step. The CDC is implemented with various radiation hardened-by-design techniques to withstand TID-induced performance degradation and equipped with TMR to minimize SEE-induced fluctuations. The performance of the CDC prototype under heavy-ion exposure is also evaluated in this experiment. The results are analyzed to determine the susceptibility of SEE-induced fluctuations and validate the effectiveness of the various radiation-hardened design techniques used inside the chip. This chapter is arranged as follows: Section 5.1 discusses the prior art. Section 5.2 elaborates on the architecture and explains the proposed time-based $\Delta\Sigma$-CDC theory. Section 5.3 discusses the circuit-level

Parts of this chapter were adapted from two papers published in the IEEE LASCAS and ISCAS 2021 conference proceedings [145, 146].

implementation of the CDC in detail. The measurement setup for the CDC design is described in Sect. 5.4, and the results obtained are analyzed and discussed to assess the performance. Section 5.4 also describes in detail the experimental setup used for the heavy-ion experiment and presents the radiation test results. Section 5.5 concludes the chapter by summarizing the performance parameters of the implemented $\Delta\Sigma$-CDC prototype samples.

5.1   Introduction

Capacitance sensors are increasingly being used for the measurement of physical quantities like pressure, humidity [147], displacement, as well as several chemical and biological parameters. An integrated solution requires an electronic interface to vary the capacitance as per the sensing parameter, followed by a CDC to measure the variation [148]. CDCs generally avoid static power consumption and are well-suited for energy-constrained applications. Applications like accelerometers [149], gyroscopes [150], and humidity sensors require CDCs to have sub-Femto-farad resolution with measurement time in the order of milliseconds. However, few applications (in-flow capacitance sensing, airborne particle detectors, etc. [148]) require high-speed conversion and precision measurement.

   Typically, in voltage-domain signal processing, the capacitance value is converted to a voltage or current quantity and measured using an ADC. However, it can also be converted to frequency- or time-based information and be measured using TDCs. Compared to their voltage-domain counterparts, time-based CDCs [151] implemented in modern CMOS technologies are semi-digital, do not require power-hungry OTAs, and their quantization capability (time resolution) enjoys the downscaling profile. Oversampling, quantization noise-shaped CDCs trade-off higher resolution for decreased energy efficiency. Time-based $\Delta\Sigma$ modulators include VCOs to integrate time-based information in the phase domain to achieve that. However, such implementations get affected by the intrinsic VCO non-linearity and added jitter from the oscillators. Alternatively, the loop filters [90] inside time-based $\Delta\Sigma$ modulators can be implemented using discrete-time charge-based time-based integrators, which provide better linearity and lower noise contribution.

5.2   Proposed Time-based Delta-Sigma CDC

Figure 5.1 illustrates the overview of the proposed time-based CDC, which works based on the principle of a first-order $\Delta\Sigma$ modulator. It performs a ratiometric measurement of the sensing capacitor C_S with respect to the full-scale value of capacitor C_{FS}. A time delay amounting to T_S, between the rising transitions of the clock edge and the output, is generated by the capacitance C_S, which is then compared to the reference time delay T_{REF}. Instead of using VTCs with a voltage-to-time conversion technique, a method of direct capacitance-to-time conversion is utilized here. Usually, VTCs suffer from non-linearity at the output

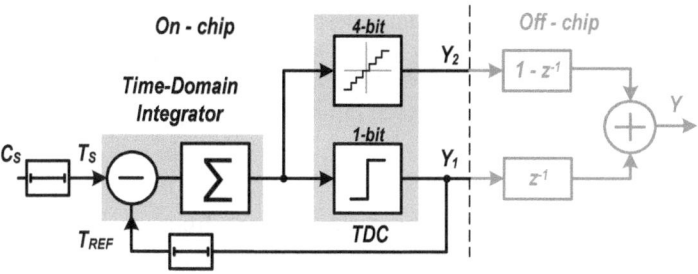

Fig. 5.1 Details of the proposed time-based $\Delta\Sigma$-CDC

and appear to be a significant limiting factor for achieving high ENOB while included with time-based circuits. Considering this, the time delays are generated directly from the sensing and reference capacitors to help reduce the non-linearity at the sensor input terminals.

The half-delay time-based integrator (TI) computes the difference between the delays T_S and T_{REF} and accumulates the time-based error e_T, which can be computed as $e_T[n] = T_S[n] - T_{REF}[n]$ for every clock cycle "n". A single-bit time-based comparator quantizes the accumulated error ($\sum e_T$). It generates output $Y_1 = 1$ and 0 for positive and negative error information, respectively. The density of occurrences of logic-1 and logic-0 at the output digital bit-stream Y_1 can be computed to be proportional to sensing capacitance C_S.

The single-bit output Y_1 provides direct feedback to the T_{REF} generating circuit, which could switch between T_{FS} and T_o for the value of $Y_1 = 1$ and 0 respectively. The time-based integrator acts to minimize the accumulated error over time using the closed negative feedback loop. Now, if the integrator takes "M" clock cycles to minimize $\sum e_T$ such that $\sum e_T \to 0$, and the averaged output digital bit-stream $\overline{Y_1}$ for "M" clock cycles can be computed as

$$M.T_S = M.\overline{Y_1}T_{FS} + M.(1 - \overline{Y_1})T_o \qquad (5.1)$$

Similar to the dual quantization-based approach used in [147], a second multi-bit TDC with higher resolution is utilized to improve the overall noise performance in this implementation. The second TDC is used with the single-bit one to quantize the integrated time-based error and produce a 4-bit digital output Y_2. The outputs Y_1 and Y_2 from those two course 1-bit and fine 4-bit TDCs are available at the CDC's output terminals and combined using an off-chip discrete-time filter to produce the final digital output Y.

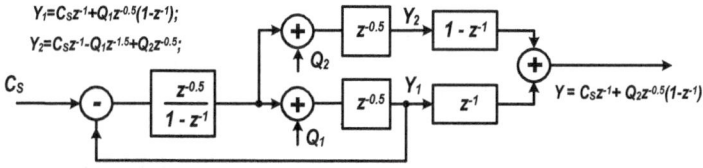

Fig. 5.2 Theoretical model of the proposed time-based $\Delta\Sigma$-CDC

Figure 5.2 shows the theoretical model of the proposed CDC with the signal flow diagram. As illustrated in Fig. 5.2, the output Y_1 and Y_2 can be derived as

$$Y_1 = C_S z^{-1} + Q_1 z^{-0.5}(1 - z^{-1}) \tag{5.2}$$

$$Y_2 = C_S z^{-1} - Q_1 z^{-1.5} + Q_2 z^{-0.5} \tag{5.3}$$

where Q_1 and Q_2 are the quantization errors resulting from the 1-bit and 4-bit TDCs, respectively. The combined filtered output Y can be derived using Eqs. 5.2 and 5.3 and can be expressed as

$$Y = Y_1 z^{-1} + Y_2(1 - z^{-1}) = C_S z^{-1} + Q_2 z^{-0.5}(1 - z^{-1}) \tag{5.4}$$

The final filtered output Y contains only the quantization error Q_2, much lower than Q_1. Thus the proposed $\Delta\Sigma$-CDC produces a first-order noise-shaped digital output with much less quantization noise similar to its multi-bit counterparts. However, unlike them, the proposed CDC does not include multi-bit feedback and can be implemented with less design complexity.

5.3   Circuit Implementation

The details of the circuit-level implementations of the complete $\Delta\Sigma$-CDC architecture are illustrated in Fig. 5.3.

5.3.1   Sensor and Reference Delay

As shown in Fig. 5.3a, the sensing capacitance $C_S \in [0, C_{FS}]$ generates a proportional time delay $T_S \in [T_o, T_{FS}]$. At each rising edge of CLK, T_{clr} discharges the capacitors $(C_{par} + C_S)$. Afterward, the capacitors start to charge exponentially. When the voltage crosses the switching threshold (V_{TH}) of the output buffer, it generates a rising edge with delay, $T_S = k_R(C_{par} + C_S)$, where $k_R = R \ln \frac{V_{DD}}{V_{DD} - V_{TH}}$. Similarly, the reference delay cell produces a rising edge with delay T_{REF}, controlled by the feedback generated from the output $Y_1(= D_7)$. T_{REF} switches between $T_{FS} = k_R(C_{par} + C_{FS})$ and $T_o = k_R C_{par}$ fro the values of $D_7 = 1$ and 0, respectively.

The delayed pair of rising transitions T_S and T_{REF} is directly connected to the time-difference detector (TD). Similar to a phase-frequency detector, the TD comprises two D-flipflops with an AND gate with T_d gate delay driving the RESET node in feedback. When the rising edge T_S arrives earlier than T_{REF} such that time delay $T_{REF} > T_S$, the TD generates at its outputs two pulse width with values equal to $(T_{REF} - T_S + T_d)$ and T_d,

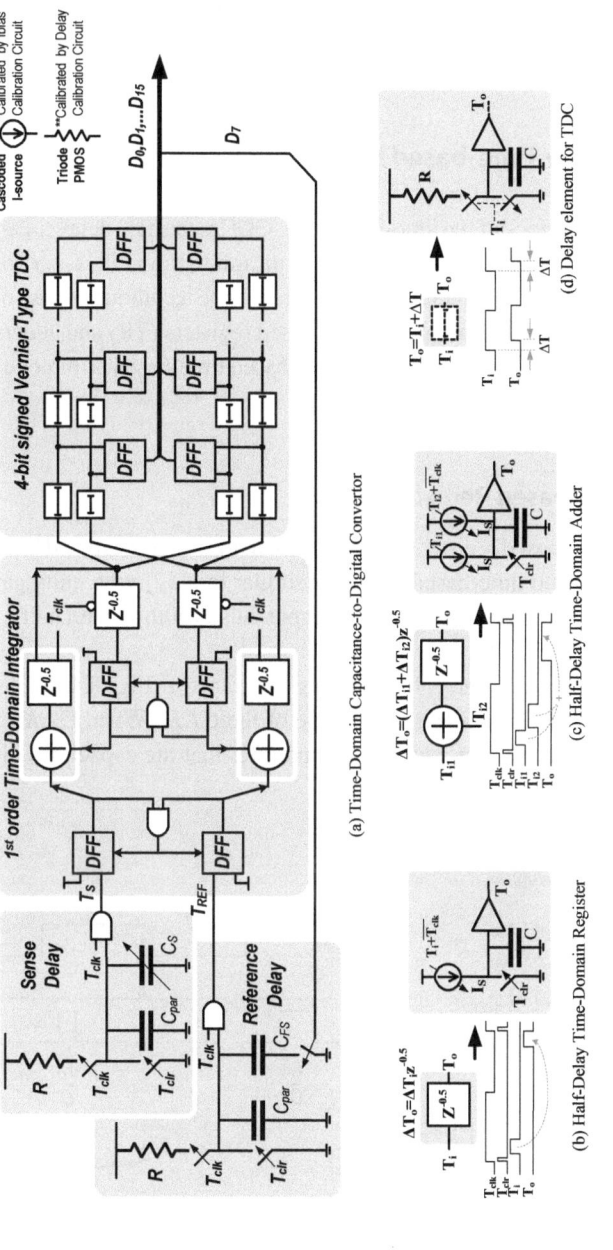

(a) Time-Domain Capacitance-to-Digital Convertor

(b) Half-Delay Time-Domain Register

(c) Half-Delay Time-Domain Adder

(d) Delay element for TDC

Fig. 5.3 Circuit implementation of the time-based $\Delta\Sigma$-CDC

respectively. Otherwise, if T_S arrives late than T_{REF} such that the delay $T_{REF} < T_S$ the TD generates two pulse widths with the values T_d and $(T_S - T_{REF} + T_d)$, respectively. Now, the averaged output $\overline{Y}_1$ from Eq. 5.1, can be expressed as a ratiometric quantity such that,

$$\overline{Y}_1 = \frac{T_S - T_o}{T_{FS} - T_o} = \frac{C_S}{C_{FS}} \tag{5.5}$$

5.3.2 Half-delay Time-based Integrator

The critical building block of the time-based $\Delta\Sigma$-CDC is the half-delay time-based integrator (TI). It accumulates the time difference between the time delays (T_S, T_{REF}) over the sampling clock cycles, following the principle of the discrete-time technique of first-order integration. The integrator is built using half-delay time-based registers (TR) and adders (TA). The TRs and TAs' working principle is based on charge-based time-based arithmetic units, which are used in [89, 145, 152].

Half-delay Time-based Register

The half-delay time-based register (TR) as shown in Fig. 5.4 is designed based on a constant-current charging-based time-based amplifier similar to [152] with unity gain, such that the output pulse-width T_o is generated from the input pulse width T_i with half-clock cycle delay $z^{-0.5}$ such that $T_o = T_i z^{-0.5}$.

As illustrated with the help of the timing diagrams in Fig. 5.4, T_{CLR} discharges the capacitor C inside the TR at each rising edge of the CLK. When $CLK=1$, the current I_1 charges C for the time period equal to T_i, and after that the capacitor C holds the charge deposited, which can be expressed as

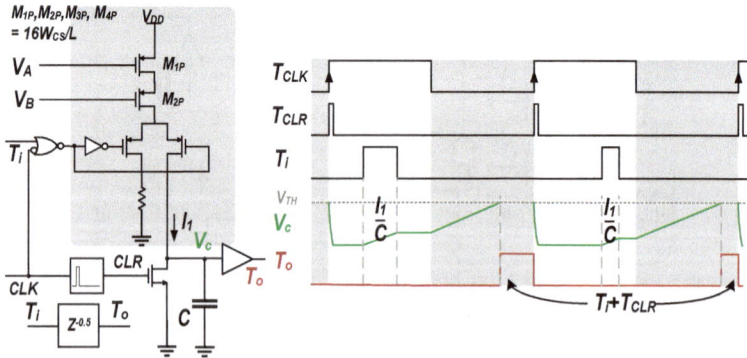

Fig. 5.4 A schematic representation of the circuit and the timing diagram of the half-delay time-based register

$$\Delta Q_H = I_1 T_i \tag{5.6}$$

When CLK goes low, I_1 starts again to charge the capacitor C. Now the digital buffer at the output node of C switches when the voltage across C, V_C crosses V_{TH}, the switching threshold voltage of the buffer, and produces a pulse width equal to T_o. The charge stored during $CLK=0$ during the time period $T_{CLK}/2 - T_o$ till V_C crosses V_{TH} can be expressed as

$$\Delta Q_L = I_1(T_{CLK}/2 - T_o) \tag{5.7}$$

The maximum input range for the TR, which can be considered the full-scale value, equals $T_{CLK}/2 - T_{CLR}$. Now, the value of the current I_1 is set such that it takes the full-scale value to charge the capacitor C from 0 V to V_{TH}. Then it can be derived that

$$(I_1/C)(T_{CLK}/2 - T_{CLR}) = V_{TH} \tag{5.8}$$

Now for each clock cycle, the total charge stored on C till V_C crosses V_{TH} can be derived as

$$C \cdot V_{TH} = \Delta Q_H z^{-0.5} + \Delta Q_L \tag{5.9}$$

Then the output T_o in the single-ended architecture as in this implementation can be computed using Eqs. 5.6, 5.7, 5.8 and 5.9 and can be expressed as

$$T_o = T_i z^{-0.5} + T_{CLR} \tag{5.10}$$

Now, in the pseudo-differential architecture case, each half is made of such a TR circuit. Then the differential output is defined as the time difference between the pulse width at the output terminals. In other words, the time difference between the rising and falling edges is synchronized with the clock transitions. Then the differential output, ΔT_o can be expressed as $\Delta T_o = \Delta T_i z^{-0.5}$. Although ΔT_o is not influenced by the common-mode time-based signal (T_{CLR}), it limits the maximum allowable input time difference amounting to $(T_{CLK}/2 - T_{CLR})$. For the operation of the TR, the minimum value of the current $I_{1,min}$ can be derived as $I_{1,min} = CV_{TH}/(\frac{T_{CLK}}{2} - T_{CLR})$. As implemented in this design the current $I_{1,min}$ equals to 15 μA for T_{CLK} = 10 ns, C = 120 fF, T_{CLR} = 0.2 ns and $V_{TH} = V_{DD}/2 = $ 0.6 V.

Half-delay Time-based Adder

The 2-input single-ended half-delay time-based adder (TA) as shown in Fig. 5.5 performs the summation of two pulse-widths T_{i1}, T_{i2} at the input terminals. It produces a pulse-width equal to T_o at the output with a half-clock cycle delay. As depicted in Fig. 5.5, when CLK goes high, the capacitor C can be charged using two matched cascode current sources I_1 and I_2. The currents are switched by the input pulse widths T_{i1} and T_{i2}, respectively. Now based on the occurrence T_{i1} and T_{i2}, there is the possibility of two use cases during the "sampling" phase as depicted in Fig. 5.5.

- For case (i), when the time period T_{i1} overlaps with T_{i2}, the total charge ΔQ_H stored in C during sampling can be expressed as

$$\Delta Q_H = I_1(T_{i1} - T_{ov}) + (I_1 + I_2)T_{ov} + I_2(T_{i2} - T_{ov}) \tag{5.11}$$

- For case (ii), if there is no overlap between the time periods T_{i1} and T_{i2}, the total charge ΔQ_H stored in C can be expressed as

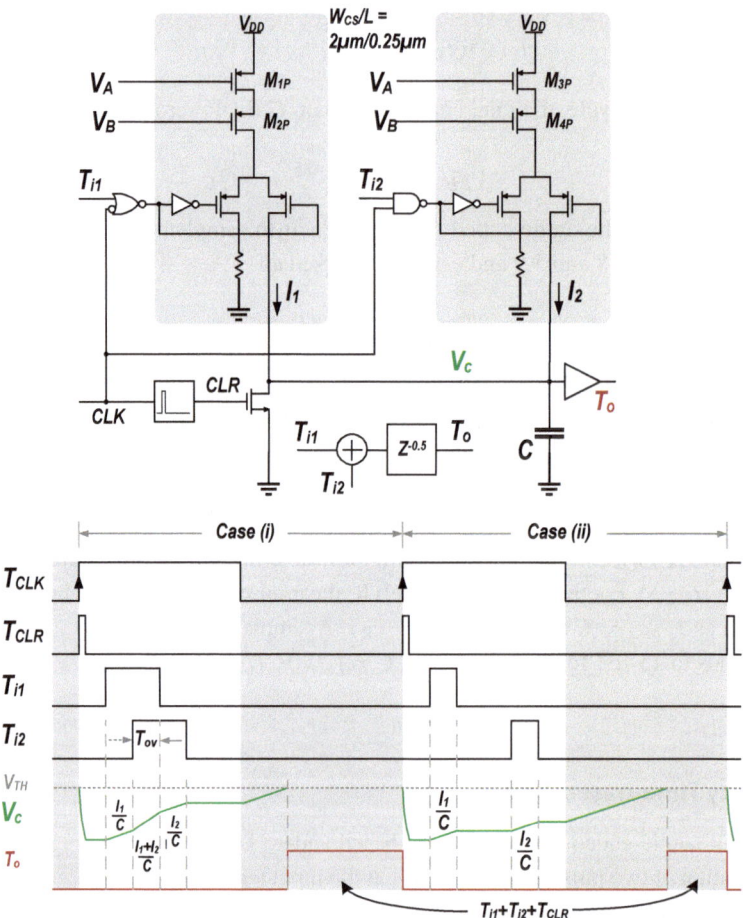

Fig. 5.5 A schematic representation of the circuit and the timing diagrams of the half-delay time-based adder

$$\Delta Q_H = I_1 T_{i1} + I_2 T_{i2} \qquad (5.12)$$

Now when CLK becomes low, only I_1 continues to charge the capacitor C till V_C crosses V_{TH}, and then the total charge stored in that time period equals $\Delta Q_L = I_1(T_{CLK}/2 - T_o)$.

Similar to the approach in computing the output pulse width T_o as in Eq. 5.10, the output pulse width can also be derived for both cases. The output T_o for the TA based on single-ended implementation can be expressed as

$$T_o = (T_{i1} + k T_{i2}) z^{-0.5} + T_{CLR} \qquad (5.13)$$

where k represents the current ratio I_2/I_1. For matched current sources ($I_1 = I_2$) as in this implementation, Eq. 5.13 can be simplified as

$$T_o = (T_{i1} + T_{i2}) z^{-0.5} + T_{CLR} \qquad (5.14)$$

Now, in a pseudo-differential architecture with matched current sources ($I_1 = I_2$) the differential output can be expressed as

$$\Delta T_o = (\Delta T_{i1} + \Delta T_{i2}) z^{-0.5} \qquad (5.15)$$

Time-difference Detector

The working principle of the time-difference detector (TD) is shown in Fig. 5.6a. It closely follows the operation of a phase-frequency detector (PFD) circuit. It comprises two D-flipflops with an AND gate driving the RESET node in feedback. The states of the TD circuit follow the transitions as depicted in Fig. 5.6b. The TD produces two pulse widths proportional to the time differences between the rising edges of the input signals, as illustrated in Fig. 5.6c. The AND-gate delay (τ_d) in feedback ensures a minimum width (τ_d) pulse at the outputs, which in turn improves the sensitivity of the next stage toward small ($< 50 ps$) input time differences.

As shown in Fig. 5.3a, the TD is used at the input terminals of TA. It helps to generate two pulse widths, the difference of which is proportional to the time differences between the rising edges of the input time-based signals. The TR passes on the outputs of the TA output with a half-clock cycle delay to the next stage TDC. The TR also provides feedback to the TA using the TD. In a fully differential design, as in this implementation, the overall z-domain system transfer function of the TI can be expressed as

$$H(z) = z^{-0.5}/(1 - z^{-1}) \qquad (5.16)$$

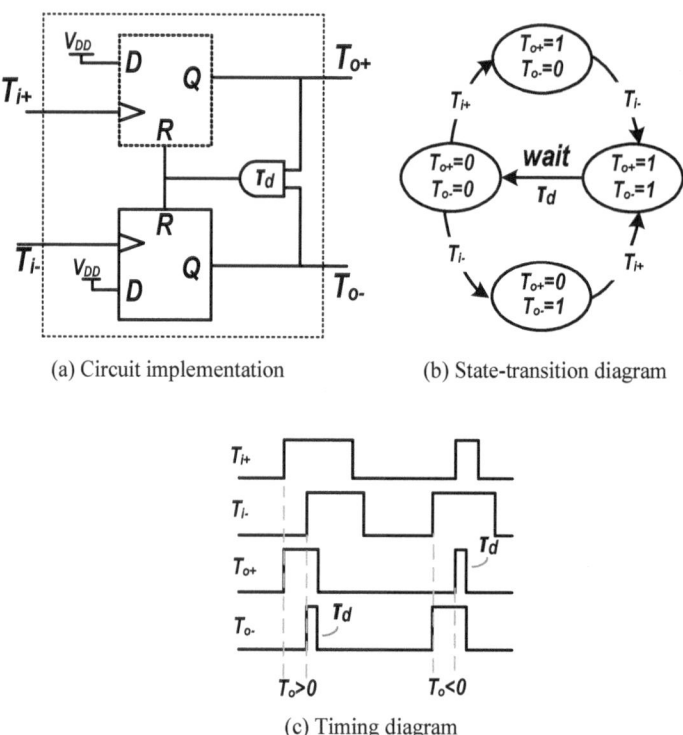

(a) Circuit implementation   (b) State-transition diagram

(c) Timing diagram

Fig. 5.6 Overview of the time-difference detector (TD)

5.3.3 Time-to-Digital Converter

In the dual quantization-based approach (see Fig. 5.2) explained in Sect. 5.2, two TDCs with different quantization levels are used inside the $\Delta\Sigma$ modulator arranged in a MASH 1–0 configuration. However, the TDCs are combined in a single unit to reduce area and power consumption in the actual implementation. The combined unit is designed as a 4-bit signed vernier-type delay-line-based TDC, providing a time resolution of 0.5 ns with a detectable range of ± 4 ns. As seen in Fig. 5.3d, the TDC employs two different delay chains with 16 D-flipflops to provide 16-bit thermometric code $D_0...D_{15}$ at the output. The middle one, D_7, basically determines whether the time difference between the occurrence of the rising edges at the TDC input terminals is positive or negative. It directly provides single-bit feedback to the reference delay T_{REF} generating circuit. As illustrated in Sect. 5.2, the digital output $D_7(=Y_1)$ is combined with $D_0...D_{15}(=Y_2)$ using an off-chip discrete-time filter to generate the final digital output "Y".

The delay cells of the TDC, as depicted in Fig. 5.3d are implemented with "R" and "C" elements together with buffers at the outputs. If the capacitors used in the two different delay chains are C_1 and C_2, namely, the effective time resolution $T_{res} \propto (C_1 - C_2)$, where

$(C_1 > C_2)$. The identical implementation of the delay chains helps cancel out the first-order process and temperature variations of the time resolution.

5.3.4 Calibration Circuit

The charging current I_S ($= I_1, I_2$) used in the TRs and TAs needs to be adjusted such that $(I_S/C)(T_{CLK}/2 - T_{CLR}) = V_{TH}$ and it varies with changing temperature and different process conditions. In this case, the allowable range of the input time difference equals to $\pm(T_{CLK}/2 - T_{CLR})$, which is limited by the variation on V_{TH} and I_S. A DLL-based foreground calibration is used to adjust and compensate for the variations. As shown in Fig. 5.7, the circuit comprises a half-delay TR, a 16-cell shift register, and a bias circuit with a switchable 4-bit thermometer-coded current reference. The off-chip resistor in the current bias generator is used for coarse tuning to accommodate larger variations across different process conditions.

The TR in the calibration circuit produces a rising transition after a delay of T_p following the occurrence of T_{CLR}. Now, $T_p(= C.V_{TH}/I_S)$ is compared with respect to the falling edge of the clock using a D-flipflop. The D-flipflop produces an error output 1 when the delay $T_p + T_{CLR} > T_{CLK}/2$ and vice versa. For a positive error, the shift register shifts right and enables the LSBs of the current reference to increase I_S and reduce T_p to adjust $T_p \rightarrow [T_{CLK}/2 - T_{CLR}]$. As elaborated with Fig. 5.8, it starts with an RST pulse setting the

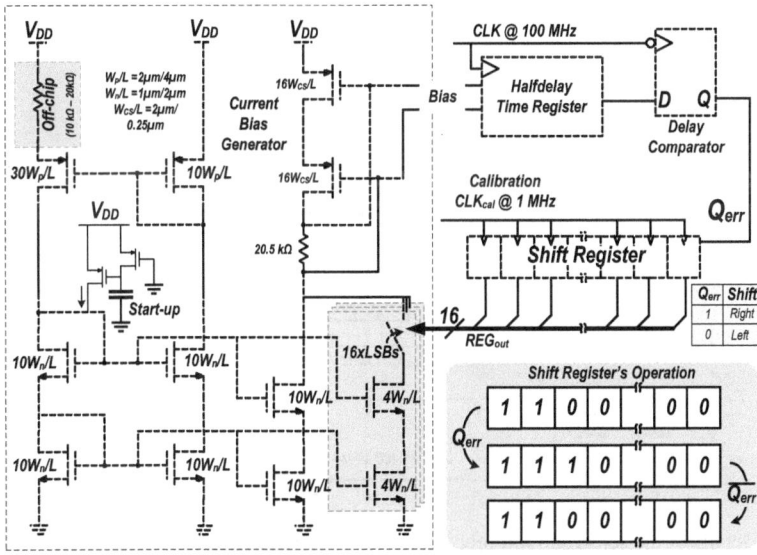

Fig. 5.7 An overview of the architecture of the calibration circuit with a detailed schematic for generating the bias current used in time-based registers and adders

output of the shift register to 8. After that, the calibration phase continues for a maximum of 8 clock cycles to adjust I_S to its desired value. It ends when the output continues to toggle for 1 LSB-bit, and therefore a maximum of $\pm1/2$ LSB error is expected after calibration.

The sense and reference delays (T_S, T_{REF}) are also sensitive to process and temperature variations. Similar to the previous method, a second DLL configures the resistor array to adjust the delays. As shown in Fig. 5.9, the output of the delay element is compared to the half-clock period $T_{CLK}/2$ using a D-flipflop. It acts as the time-based comparator and produces an error output $Q_{err} = 1$ for positive and 0 for negative errors. As per the error output Q_{err}, the 16-cell shift register enables or disables the resistors in the delay element, such that $T_{FS} \rightarrow T_{CLK}/2$. Due to the similarity in implementation, the calibration of T_S and T_{REF} also helps to compensate for process and temperature variations in the RC-delay elements of the TDC.

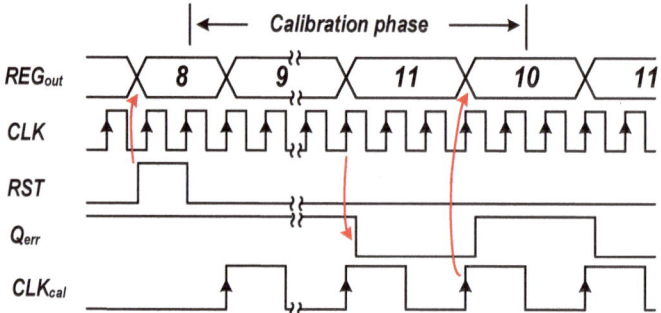

Fig. 5.8 Timing diagram of the calibration circuit

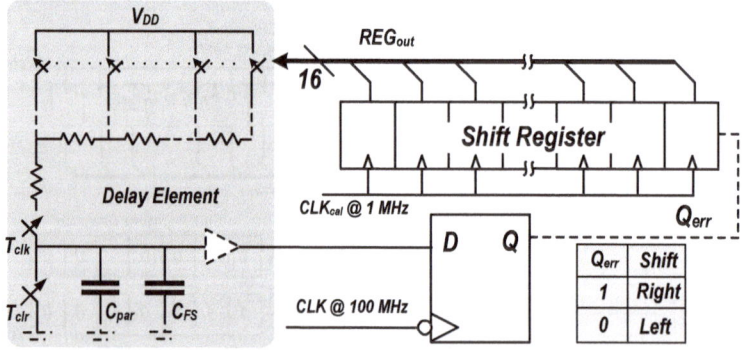

Fig. 5.9 Details of the sense and reference delay calibration circuit

5.4   Measurement Results

The proposed time-based $\Delta\Sigma$ CDC is implemented using a 65 nm CMOS Technology. Figure 5.10 shows the micrograph (2x2 mm^2) of the fabricated die and Fig. 5.11 shows details of the setup used for measurement. The CDC core circuit consumes 2 mW from a 1.2 V power supply at a 100 MHz clock frequency.

Figure 5.12a shows the estimated power distribution for different core-circuit blocks in terms of percentages of total power consumed. The prototype can measure capacitance in the range of 0–3.75 pF, and during the measurement, the on-chip capacitance array (with a similar range) is enabled by external digital inputs. Figure 5.12b shows the final combined digital output with respect to different capacitance inputs, and the response shows a good linear transfer characteristic.

The output spectrum of the CDC is shown in Fig. 5.13a. The final digital output is computed using "MATLAB" software from the digital bit stream recorded from the die sample during measurement. The frequency spectrum is computed using 2^{16}-pt FFT with a Hanning window of the same length. As shown in Fig. 5.13a, the spectrum follows a 20 dB/dec up-conversion trend and contains high-frequency tones generated from the modulation of DC input and the sampling clock. During off-chip digital signal processing using "MATLAB" software, those are filtered out using the digital decimation filter afterward. Unlike voltage-domain $\Delta\Sigma$ ADCs, the dynamic performance of the time-based $\Delta\Sigma$ modulators particularly used in CDC interfaces lacks the feasibility to be tested with sinusoidal excitation. Although MOS varactors varying with sinusoidal control voltage can be used, the

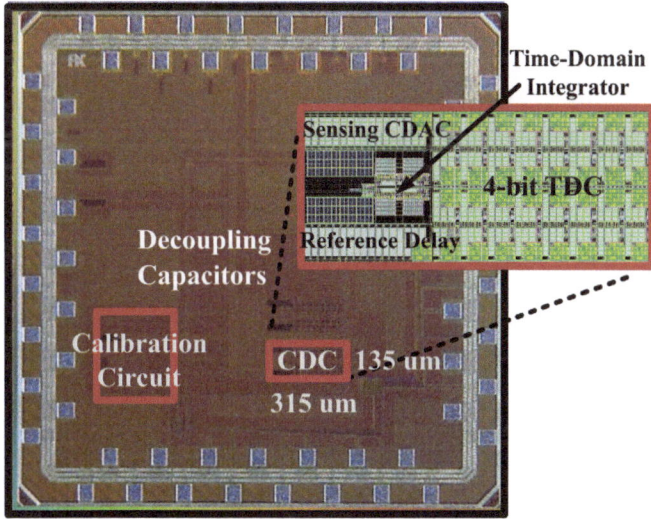

Fig. 5.10 Micrograph of the CDC prototype (2×2 mm^2)

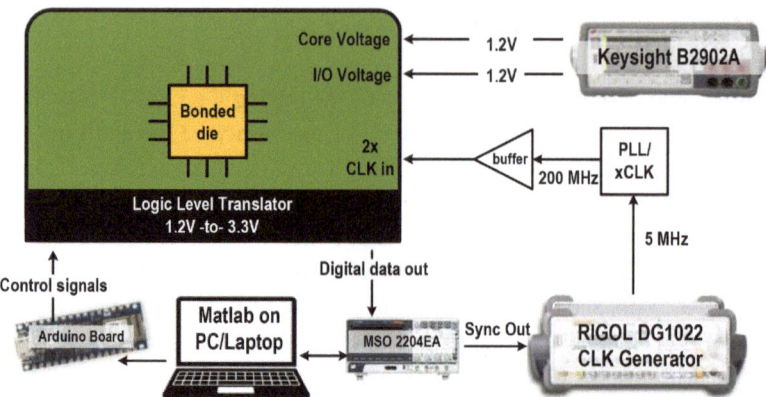

Fig. 5.11 Details of measurement setup for the $\Delta\Sigma$-CDC prototype

Fig. 5.12 **a** The breakdown of power dissipation of the core-circuit blocks and **b** the transfer characteristics of the $\Delta\Sigma$-CDC

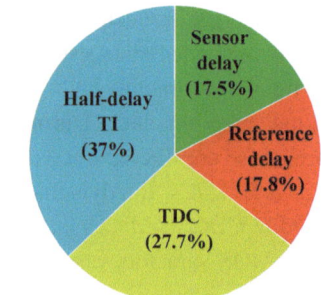

(a)

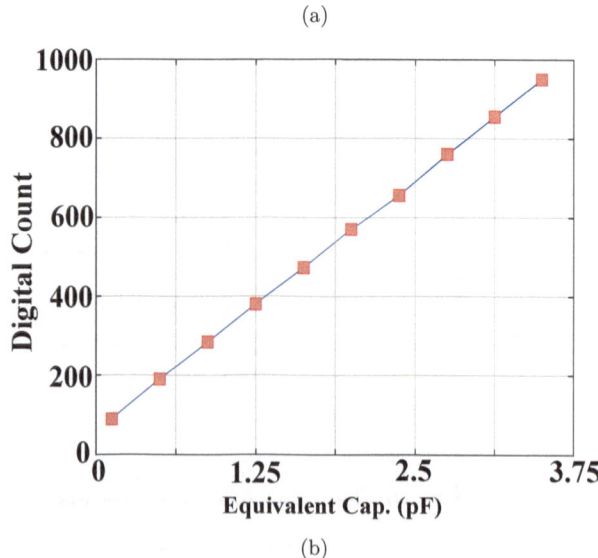

(b)

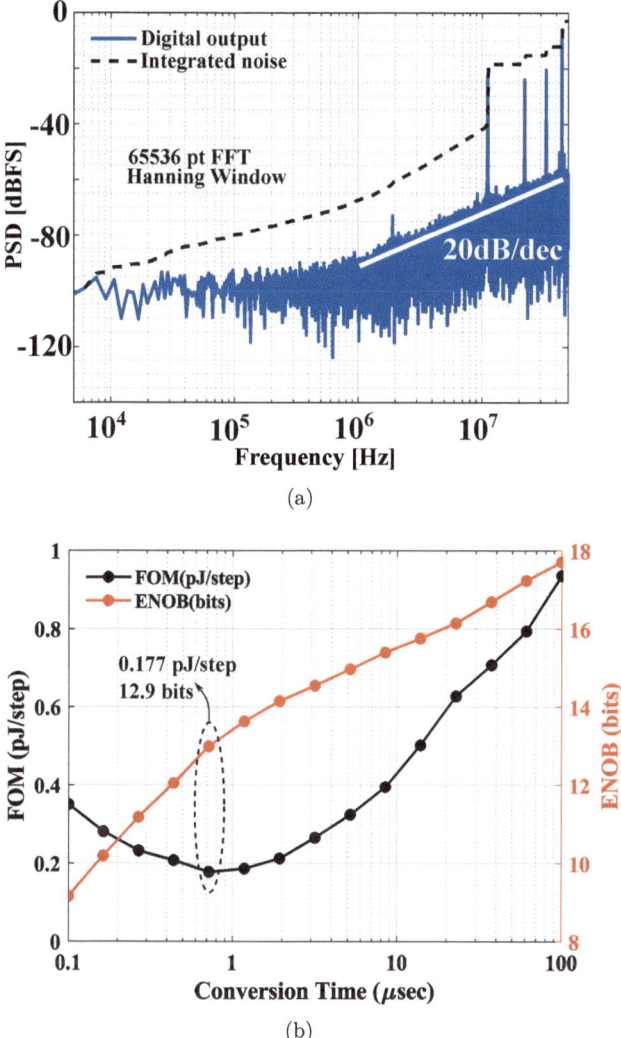

(a)

(b)

Fig. 5.13 **a** The output spectrum of the $\Delta\Sigma$-CDC and **b** the variation of FOM, ENOB with respect to conversion time

performance gets limited by the non-linearity in the interface providing sinusoidal variation to capacitors. Similar to the approach followed in [153–156], the SNR based on integrated rms noise is used for the characterization. The SNR can be expressed as

$$SNR(dB) = 20\log_{10}\left(\frac{\text{Capacitance Range}/2\sqrt{2}}{\text{Cap.Resolution}}\right)$$

where "Capacitance Range" is the full-scale measurement range and "Resolution" is the root-mean-square value of in-band integrated noise. Subsequently, the effective-number-of-bit (ENOB) based on SNR and the Walden figure-of-merit (FOM) can be defined as, ENOB = (SNR-1.76)/6.02, and FOM = Energy$/2^{ENOB}$ respectively. Figure 5.13b shows the variation of measured FOM and ENOB with respect to conversion time. At lower conversion time, the output is dominated by the first-order high-pass shaped TDC quantization noise. As the conversion time (>1 μs) increases, the thermal noise's contribution becomes dominant. The optimum FOM of 0.18 pJ/conversion step for the CDC prototype is obtained at 0.76 μs conversion time.

Performance Comparison with State-of-the-Art Designs

Table 5.1 compares the performance of the proposed CDC with respect to previous state-of-the-art designs as reported in the literature. This work accomplishes the fastest conversion time compared to its counterparts reported so far in the literature, enabling it to be utilized with sensor arrays with high throughput. Overall, this proposed CDC achieves very high ENOB ($>$ 12 bits), and while compared to only voltage-domain or time-based implementations, it achieves the best energy efficiency. In comparison to other hybrid implementations incorporating both time-based as well as voltage-domain techniques, this work appears third in line in terms of spent energy per conversion steps.

Table 5.1 Performance comparison with state-of-the-art CDCs

	JSSC-20 [151]	JSSC-17 [153]	**This work**	TIM-19 [154]	TCASI-18 [155]	JSSC-19 [147]	ISSCC-15 [156]
Technology (nm)	40	40	**65**	180	180	180	160
Method	Time + Voltage	Time + Voltage	**Time**	Voltage	Voltage	Voltage	Time
Topology	$\Delta\Sigma$+SAR	SAR+VCO	$\Delta\Sigma$	SAR	$\Delta\Sigma$	$\Delta\Sigma$+SAR	PM†
Power supply (V)	0.6/1.1	1	**1.2**	0.8/1	1.5	1.1	1
Cap. range (pF)	0–5	0–5	**0–3.75**	0–3.6	1–10^5	0–18.12	0–8
Conv. time (μs)	12.5	1	**0.76**	810	128	850	210
Energy (nJ)	0.083	0.075	**1.52**	1.28	1.92	2.62	2.902
ENOB(bits)	12.3	10.4	**12.9**	12.7	13	11.8	10.6
SNR (dB)	75.8	64.2	**79.4**	78.5	80	72.8	65.6
FOM (fJ/conv.step)	16	55	**177**	187	230	660	1870

†PM = Pulse-width Modulation

5.4.1 Heavy-Ion Experiment

The performance of the CDC prototypes was evaluated under heavy-ion exposure to assess the SEL sensitivity in the RADEF facility at the University of Jyvaskyla, Finland. The test setup for the radiation experiment included three PCBs: "sensor" PCB, "control" PCB, and "FPGA-interface" PCB. The die samples were mounted on the "sensor" PCB consisting of 1.2–3.3 V logic-level translators and LVDS transceivers and placed inside the irradiation chamber under the beam target area. The "control" PCB consisted of another set of LVDS transceivers and LDOs, which were used to send control signals, power supply, and read back digitized sensor output from the sensor PCB. The "FPGA-interface" PCB was used to send digital configurations and capture the sensor output bit-stream through the "control" PCB. A PC/laptop was kept in the control room, and it used an active USB-A cable of 10 m length to communicate to the "FPGA-interface" PCB. During the heavy-ion experiment, three units of CDC samples were irradiated at ambient temperature in the air with Xe-ion for 3 different V_{DD} settings: 1.08 V, 1.2 V, and 1.32 V. The CDC designs operating at each of the settings were irradiated for approximately 12–16 minutes to achieve a fluence of at least $5 \times 10^7\ n_{eq}/\mathrm{cm}^2$. The ion flux was configured around $7 \times 10^4\ n_{eq}/\mathrm{cm}^2/\mathrm{s}$. The Xe-ion beam energy was set at 2059 MeV to obtain a LET of approximately 65 MeV.cm^2/mg. The power supply currents for the CDC core and input-output interfaces are monitored during the irradiation to observe possible events due to SEEs (e.g., SEU, SEL). A protection limit of 10 mA was set on all supply currents, resulting in a power-down state if any readings crossed the specified limit due to an SEL-induced current surge. During the experiment, no such event of SEL was observed, translating to an SEL cross section below 2 μm^2. However, a few minor fluctuations in supply current ($>=$ 1 mA) were observed, which were considered to happen due to SEUs in the circuit.

5.5 Conclusion

This chapter has presented the proposed first-order $\Delta\Sigma$ noise-shaped CDC, which is entirely designed using time-based circuits and is highly digital. This chapter discussed in detail the CDC architecture and the measurement results. The prototype is implemented in a 65 nm CMOS Technology, leverages the merits of the higher time-based quantization capability of TDCs, and achieves an ENOB of 12.9 bits with 0.18 pJ/conversion-step energy efficiency. It consumes 2 mW from a 1.2 V supply while sampling at a 100 MHz clock frequency and supports capacitance measurement within the range of 0–3.75 pF. The performance of three different samples of CDC prototypes was tested under Xe-ion (65 MeV.cm^2/mg) exposure till it reached the fluence of $5 \times 10^7\ n_{eq}/\mathrm{cm}^2$ with a flux of $7 \times 10^4\ n_{eq}/\mathrm{cm}^2/\mathrm{s}$ to assess the susceptibility toward SEL events. As measured no SEL events (current $>=$ 10 mA) except some SEU events (current $>=$ 1 mA) are observed. This establishes the immunity of the implemented prototypes against latch-up events for ion energy $<=$65 MeV.cm^2/mg in the air at ambient temperature.

Multi-Order Differential Delta-Sigma TDC

6

Abstract

This chapter presents a novel differential $\Delta\Sigma$ time-to-digital converter implemented in a 65 nm CMOS process. The 0.018 mm^2 $\Delta\Sigma$-TDC with a maximum T_{range} of 98 ($= \pm49$) ns uses a time-based FIR filter in feedback to realize first, second, and third orders of $\Delta\Sigma$ modulation. It consumes 11–15.8 μW from a 0.6 V supply and achieves the best state-of-the-art energy efficiency of 15.6–89.2 fJ/conversion step at 10 MHz clock frequency for multiple orders (≤3) of noise-shaping filters. This chapter is arranged as follows: Section 6.1 discusses the prior art. Section 6.2 elaborates on the architecture and explains the working principle of the proposed time-based $\Delta\Sigma$-TDC. Section 6.3 discusses the circuit-level implementation of the TDC in detail. The measurement setup for the TDC design is described in Sect. 6.4, and the results obtained are also analyzed and discussed to assess the performance. Section 6.5 concludes the chapter by summarizing the performance parameters of the implemented $\Delta\Sigma$-TDC.

6.1   Introduction

Oversampling $\Delta\Sigma$-type data converters trade-off energy efficiency to improve resolution. Unlike the voltage-domain circuits, the time-based $\Delta\Sigma$-TDCs without static-power consumption suit energy-constrained interfaces and exploit the down-scaling profile with improved time resolution despite having reduced voltage headroom. TDCs are essential for all-digital PLLs [91, 157–159] and are increasingly being adapted for energy-constrained sensor readouts in scaled CMOS processes. $\Delta\Sigma$-TDCs relying on ring-oscillators (ROs) suffer from non-linear voltage-to-frequency conversion. As shown in Fig. 6.1, the $\Delta\Sigma$-TDCs implemented based on switched-ring-oscillator [157] provide a large dynamic range

© The Author(s), under exclusive license to Springer Nature Switzerland AG 2024   83
A. Karmakar et al., *Integrated Time-Based Signal Processing Circuits for Harsh
Radiation Environments*, Synthesis Lectures on Engineering, Science, and Technology,
https://doi.org/10.1007/978-3-031-40620-1_6

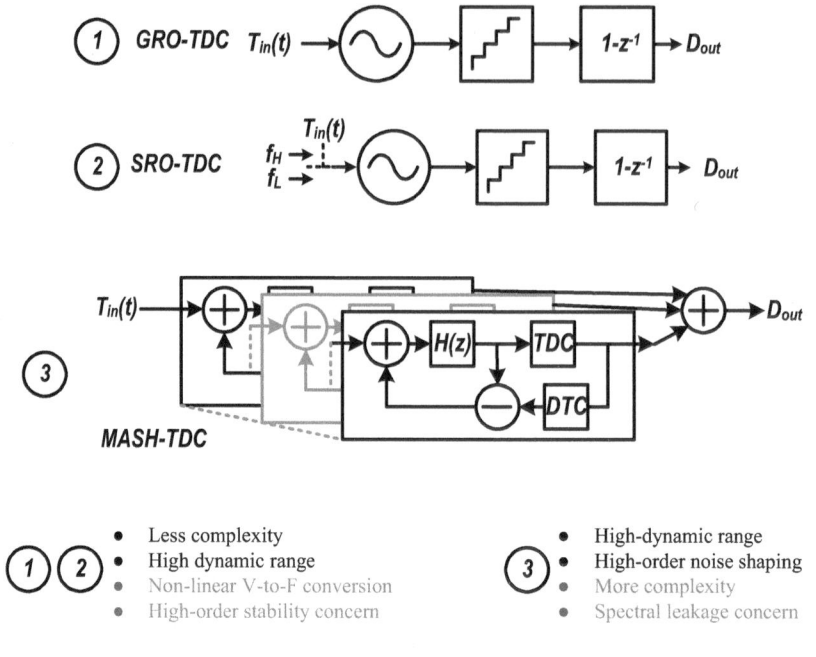

Fig. 6.1 A comparative overview of pros and cons of the proposed $\Delta\Sigma$-TDC in comparison with prior art

but are prone to higher-order stability issues. These issues are alleviated with gated-ring-oscillators [158] with high oversampling ratios, and those can also be integrated within higher-order MASH loops. However, noise-shaping TDCs with MASH architectures [91] are complex, dissipate more power, and are prone to spectral noise leakage from mismatch among cascaded branches. Alternate options to achieve higher orders of $\Delta\Sigma$ modulation are time-based discrete-time filters built around time-based arithmetic units. These energy-efficient filters utilize pulse widths or time differences as time-based signals to perform time-based signal processing. The $\Delta\Sigma$-TDC designs based on time-based IIR filters lack systematic design methodology due to fixed pole positions and suffer from stability issues in the case of higher-order implementations. The proposed $\Delta\Sigma$-TDC solves the limitation while presenting a systematic design approach at the architectural level to design a higher-order time-based discrete-time filter with configurable filter coefficients.

6.2 Proposed Architecture

The proposed $\Delta\Sigma$-TDC is designed based on the single-loop multi-order architecture of oversampling data converters using quantization error feedback. The $\Delta\Sigma$-TDC is completely implemented with time-based arithmetic units to realize time-based filters for in-band quantization noise-shaping. The order of the time-based filter is changeable using configurable filter coefficients and helps to realize three different orders: first, second, and third orders of the noise-shaping filters. An overview of the proposed architecture of the $\Delta\Sigma$-TDC is illustrated in Fig. 6.2. Overall, the proposed TDC digitizes the input time difference using oversampling techniques. It extracts the resulting quantization error (Q_e) from the output and sends it back to pre-distort the input signal to reduce the in-band quantization noise effectively. The proposed $\Delta\Sigma$-TDC is pseudo-differential; therefore, the time difference between two rising edges is used as the differential time-based input signal. The complete architecture of the $\Delta\Sigma$-TDC can be divided into two major circuit blocks in terms of design and implementation. The first block, as highlighted on the left in Fig. 6.2, can be identified with the time-based quantization error filter (QEF) built using several time-based arithmetic circuits. The second block, as highlighted on the right of the TDC's architecture shown in Fig. 6.2, can be referred to as the quantization error (Q_e) extractor circuit.

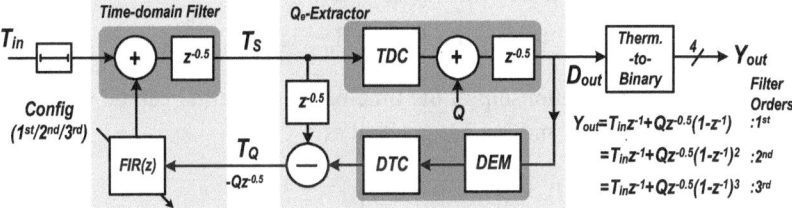

Fig. 6.2 The proposed architecture of the time-based $\Delta\Sigma$-TDC

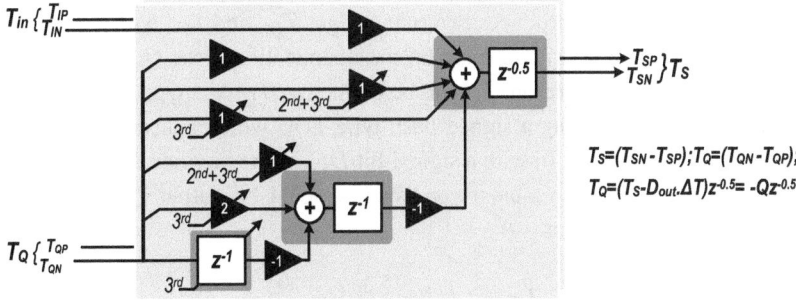

Fig. 6.3 Architecture of the time-based QEF filter used in feedback

Figure 6.3 provides a more detailed architecture of the time-based filter. The filter architecture relies on the discrete-time FIR filter implementation using time-based variables. The inputs to the filter are the time-based input signal T_{IN} and T_Q, which is proportional to the time-based quantization error (Q_e) resulting from the previous digitizations. T_{IN} can be defined as the time difference between two discrete events in terms of rising transitions: T_{INp} and T_{INn} happening in two separate PWM signals such that $T_{IN} = T_{INn} - T_{INp}$. Similarly, T_Q can be referred to as the time difference between two rising edges: T_{Qp} and T_{Qn} such that $T_Q = T_{Qn} - T_{Qp}$. When CLK is 1, the time-based filter samples the differential time-difference T_{IN} and T_Q ($\propto Q_e$). When CLK = 0, the filter generates T_S, a differential time-based signal, which can be identified as the time-difference between two rising edges at the output: T_{Sp} and T_{Sn} such that the output $T_S (= T_{Sn} - T_{Sp})$ can be computed as $T_S = T_{in}z^{-0.5} + T_Q FIR(z)z^{-0.5}$. Here, $z^{-0.5}$ and z^{-1} are used to denote half-cycle ($T_{CLK}/2$) and full-cycle (T_{CLK}) delays between the input and output. As depicted in Fig. 6.3, the filter coefficients can be adjusted by configuring (enabling or disabling) the inputs to the time-based adders, eventually leading to different orders of $\Delta\Sigma$ modulation. The transfer function of the FIR filter, FIR(z), for the different orders of $\Delta\Sigma$ modulation, can be derived as

$$\text{1st order} \rightarrow FIR(z) = 1 \tag{6.1}$$

$$\text{2nd order} \rightarrow FIR(z) = 2 - z^{-1} \tag{6.2}$$

$$\text{3rd order} \rightarrow FIR(z) = 3 - 3z^{-1} + z^{-2} \tag{6.3}$$

The overall input-output relationship of the time-based QEF filter can be written with the help of Eqs. 6.1, 6.2, and 6.3 and can be specified as

$$\text{1st order} \rightarrow T_S = T_{IN}z^{-0.5} + T_Q z^{-0.5} \tag{6.4}$$

$$\text{2nd order} \rightarrow T_S = T_{IN}z^{-0.5} + T_Q(2z^{-0.5} - z^{-1.5}) \tag{6.5}$$

$$\text{3rd order} \rightarrow T_S = T_{IN}z^{-0.5} + T_Q(3z^{-0.5} - 3z^{-1.5} + z^{-2.5}) \tag{6.6}$$

T_S, the differential time-based output of the QEF filter is connected to the Q_e extractor block. The working principle of the Q_e extractor circuit incorporating the TDC and DTC units is illustrated in Fig. 6.4, which produces two rising edges T_{Qp} and T_{Qn} with a time difference T_Q as defined earlier. Now, T_Q depends on the quantization error Q_e resulting from the previous clock cycle, and therefore T_Q can be defined as $T_Q = -Q_e z^{-0.5}$. The input time difference T_S is digitized using a signed flash-type TDC which generates a thermometric encoded 7-bit output $D_{out}[6:0]$ with a signed-bit D_{sign}. The thermometric output with the signed bit is then converted to a binary encoded signed 4-bit output $Y_{out}[3:0]$. The final output Y_{out} can be expressed as

$$Y_{out} = T_S z^{-0.5} + Q_e z^{-0.5} \tag{6.7}$$

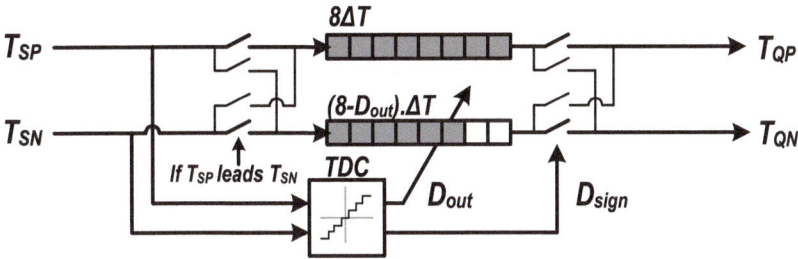

Fig. 6.4 Working principle of the Qe extractor block

With the help of an arbiter circuit, T_{START} follows the leading edge between T_{Sp} and T_{Sn}, whereas T_{STOP} follows the lagging one, such that the absolute time difference $|T_{Sp} - T_{Sn}| = |T_{START} - T_{STOP}|$. When CLK = 0, the flash-type TDC in the forward path starts the conversion at T_{START} and stops it at T_{STOP}. As mentioned earlier, it generates the digital output $D_{out}[6:0]$ proportional to the absolute time difference $|T_S| = |T_{START} - T_{STOP}|$. D_{sign} is produced based on whether $T_S > 0$ or < 0.

Now, the rising-edge T_{START} is delayed by eight delay elements (ΔT) inside the TDC such as $8\Delta T = T_{CLK}/2$. Similarly, the rising-edge T_{STOP}, which is used to stop conversion, is also delayed by an amount equal to $(8 - D_{out})\Delta T$. However, the time delay for the case of T_{STOP} is generated using the delay elements of the DTC such that the number of delay elements is controlled by D_{out}.

The timings of those rising edges with the delays are elaborated with details in Fig. 6.5. In the next clock cycle when CLK = 1, the time-based signals, i.e., the delayed pair of rising edges, are interchanged based on the value of D_{sign} such that the time difference between T_{Qp} and T_{Qn}, i.e., $T_Q = -Q_e z^{-0.5}$. Here, in this implementation, the quantization error, Q_e resulting from the time-to-digital conversion in the n-th clock cycle can be defined as $Q_e[n] = T_S[n] - D_{out}[n]\Delta T$. Now, considering $T_Q = -Q_e z^{-0.5}$, the output Y_{out} for three different cases: first, second, and third orders can be computed using Eqs. 6.4, 6.5, and 6.6 and can be expressed as

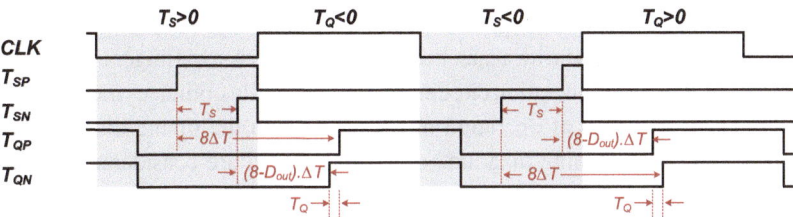

Fig. 6.5 The timing diagrams of the Qe extractor circuit

$$\text{1st order} \rightarrow Y_{out} = T_{IN}z^{-1} + Q_e z^{-0.5}(1 - z^{-1}) \tag{6.8}$$

$$\text{2nd order} \rightarrow Y_{out} = T_{IN}z^{-1} + Q_e z^{-0.5}(1 - z^{-1})^2 \tag{6.9}$$

$$\text{3rd order} \rightarrow Y_{out} = T_{IN}z^{-1} + Q_e z^{-0.5}(1 - z^{-1})^3 \tag{6.10}$$

6.3   Circuit Implementation

6.3.1   Time-based Adder

The differential QEF filter comprising the time-based FIR filter is built using three time-based adders. These adders could add multiple time-based signals (T_i) and produce a time-based output proportional to ΣT_i with a half-cycle ($z^{-0.5}$) or full-cycle (z^{-1}) delay. Here, in differential mode as this design, the time-based signals are considered as the time difference between the rising transitions. However, the pulse widths are considered time-based signals for single-ended architectures. As illustrated earlier using Fig. 6.3, the filter coefficients can be adjusted by masking or unmasking one or more inputs to the time-based adders. Each of the three time-based adders is pseudo-differential, and each half contains a pair of time-based adder units (TAUs). The working principle of TAUs can be illustrated by considering the case of a generic n-input single-ended TAU as shown in Fig. 6.6a. The TAU is designed with $n + 1$ branches of current sources, each one, when enabled, can generate an equal amount of current I_c. The TAU operates in four phases sequentially which are named "sample", "hold", "convert", and "clear". When CLK $= 1$, it samples the time-based inputs with pulse widths $T_{i[1]}$, $T_{i[2]}$,... $T_{i[n]}$ and enables the corresponding branches with current I_c to put charge on capacitor C ($= 385$ fF). At the end of the "sample" phase, it stores a charge proportional to $\Sigma(I_c T_{i[n]})$ on the capacitor C, which then holds the charge for a period of $T_{CLK}/2$. During the "convert" phase, another branch generating a current equal to I_c starts charging C again. As the voltage on C, identified as V_c crosses the switching threshold, $V_{TH}(= 0.3$ V) of the AND gate, a pulse width $T_o(\propto \Sigma T_{i[n]})$ is generated at the output. Next, in the "clear" phase, capacitor C is discharged.

The constant currents of amount I_c are produced by low-V_{th} PMOS devices, which are biased with $V_{SG} = V_{DD}$. As illustrated on the right side of Fig. 6.6a, during time-based addition, these devices remain in saturation and maintain constant I_c till V_c reaches V_{TH}. The devices are implemented with a matched layout with interleaved fingers, and it eliminates the need for separate bias generation circuits, which eventually optimizes the area and power.

The complete sequence of operation of the TAUs with the timing diagrams is shown in Fig. 6.6b. There, the timing diagrams of a 2-input TAU are considered for illustration purposes. From the timing and sequence of operation, it could be inferred further that a single TAU misses each alternate sample during the "convert" phase. Therefore, it would require a pair of TAUs to collect the samples alternatingly such that when one is operating in the "convert" phase, the other is operating in the "sample" phase and vice versa.

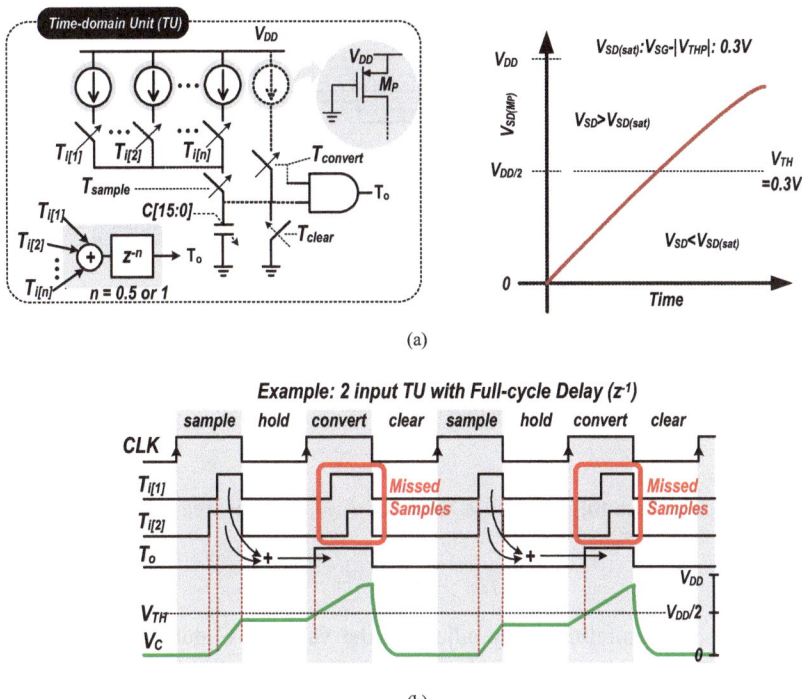

(a)

(b)

Fig. 6.6 **a** Schematic of a generic n-input single-ended time-based adder unit (TAU) and **b** the timing diagrams for a 2-input implementation

As depicted in Fig. 6.7, the time-based adder circuit uses two TAU elements operating alternatingly to perform time-based addition in the case of a single-ended configuration. In the differential configuration, the time-based adders contain two halves which require two pairs of TAU elements working in an interleaved manner. Now, in single-ended mode, there also exists T_{offset} at the output such that $T_o = [T_{i[1]}, T_{i[2]}, ... T_{i[n]}] z^{-n} + T_{offset}$, where $n = 0.5, 1$. Now, T_{offset} can be derived as the value of time-based output for zero-input pulse widths. In the case of zero inputs, i.e., for $T_{i[n]} = 0$ for $n = 1, 2, ... n$, V_c starts from 0 V in the "convert" phase and rises linearly with a slope of $\lambda = I_c / C$. As V_c crosses V_{TH},

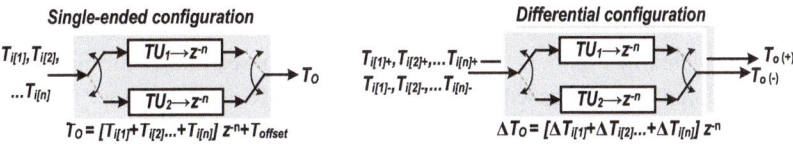

Fig. 6.7 Details of the single-ended and differential configuration of time-based adders built using TAU elements

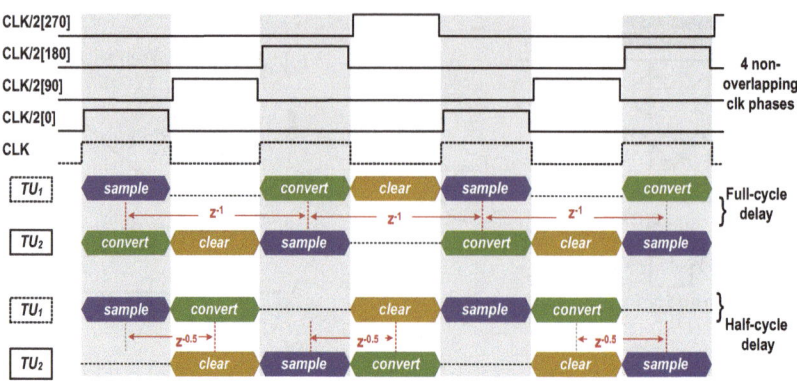

Fig. 6.8 Sequence of operation of TAUs to realize $z^{-0.5}$ or z^{-1} delay

the output switches, resulting in a pulse width of T_{offset}. Thus, the value of T_{offset} can be derived as

$$T_{offset} = \frac{T_{CLK}}{2} - \frac{V_{TH}}{\lambda} \tag{6.11}$$

T_{offset} (< 0.5 ns) is minimized by adjusting the value of C during calibration. In the differential mode of the time-based adders, as in this implementation, T_{offset} is canceled out as differential time-based signals are derived as the difference between pulse widths, i.e., the time difference between the rising transitions.

The QEF filter, as illustrated in Fig. 6.3 requires one time-based adder providing output with half-cycle ($z^{-0.5}$) delay and two time-based adders providing output with full-cycle (z^{-1}) delay. The same TAU elements can be operated with different combinations of the clock phases to realize such delays. As depicted in Fig. 6.8, the TAU elements produce a delay of z^{-1} at the output if the "hold" phase is used between the "sample" and "convert" phases. Alternatively, if the "convert" phase starts right after the "sample" phase and an "idle" phase is introduced after the "convert" phase, the TAU elements produce a delay of $z^{-0.5}$ at the output. The schematic of the circuit used to generate four such clock phases is provided in Fig. 6.9a. The clock phases are produced with ~25% duty cycle at the half-rate of the global clock ($f_{CLK} = 10$ MHz). These clock phases are realized as non-overlapping with the ON-period < $T_{CLK}/2$ to minimize leakage power during switching. However, it reduces the full-scale sampling range by 2% as the TAU elements cannot capture two transition events with a time difference more than the ON-period of the non-overlapping clock phases.

Now, as depicted in Fig. 6.3, three differential time-based adders are put in series to construct the complete time-based QEF filter. A time-difference detector (TD) circuit, as depicted in Fig. 6.9b, is used in between the stages, which relay the output pair of rising transitions to the next stage's inputs. The TD circuit works similarly to a phase frequency detector and produces a pair of pulse widths proportional to the time differences between the

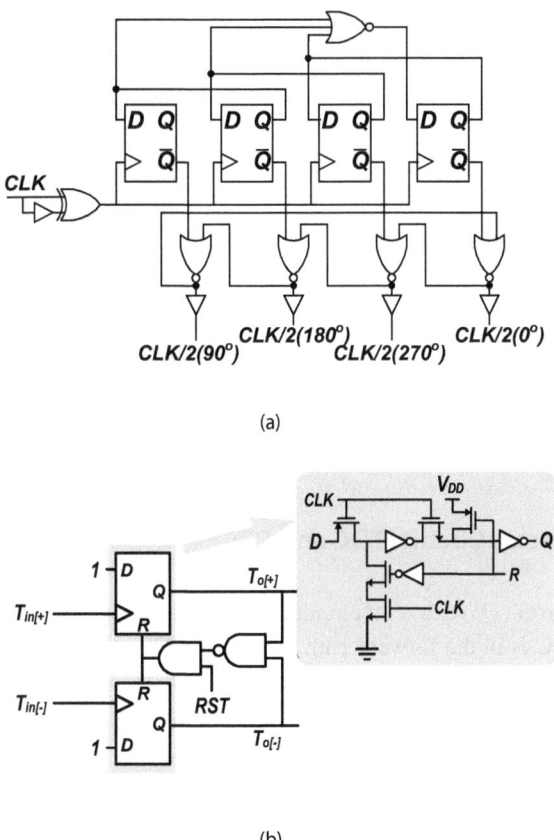

(a)

(b)

Fig. 6.9 The schematic representations of **a** the clock-phase generator circuit and **b** the time-difference detector circuit

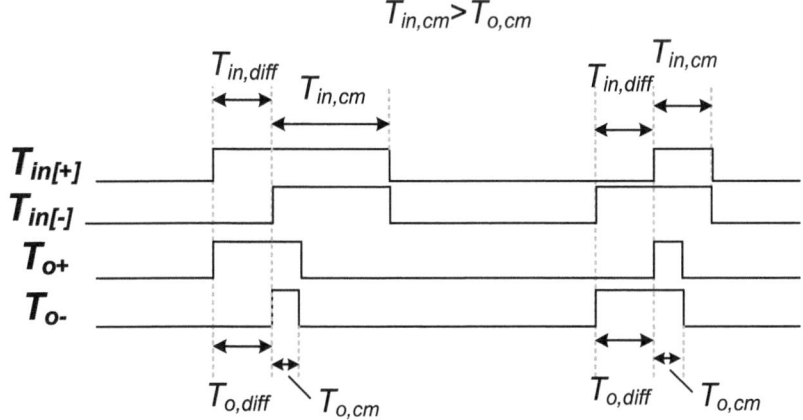

Fig. 6.10 The timing diagrams of the time-difference detector

rising transitions of the input pair. To realize negative integer gain for inter-stage time-based signals, the output differential pair of the TD circuit is interchanged so that the polarity of the time difference reverses. This could have been realized without the TD circuit, but the TD also helps decouple the "common-mode" time-based signals between the stages, similar to the "AC-coupling" technique used in voltage-based circuits. In the case of differential stages, as in this implementation, the input pair $T_{in[+]}$ and $T_{in[-]}$ are represented as PWM signals with the falling edges aligned so that the time difference between the rising-edges act as the differential time-based signal. However, in each clock cycle, the PWM signal with the minimum pulse width, i.e., $\min\{T_{in[+]}, T_{in[-]}\}$ appears as the common-mode time-based signal for the next stage's input. The TD circuit only relays the input time difference to the output with a minimum pulse width equal to the feedback path delay through the NAND gate. Thus, when used in between the stages, the TD circuit masks the common-mode time-based signal (as depicted in Fig. 6.10) regardless of its value at the input terminals.

6.3.2   Qe Extractor Circuit (TDC and DTC)

The quantization error (Q_e) extractor circuit consists of a course flash-type TDC and a DTC. The flash-type TDC is in the forward path along the $\Delta\Sigma$ loop. It digitizes the output time difference T_S, produced by the previous time-based filter. The DTC converts the digital value to an equivalent time delay, which is then subtracted from the time difference T_S to produce the resulting quantization error (Q_e). Figure 6.11 provides the overview of the circuit consisting of the forward path TDC and the DTC. The time difference T_S is bipolar and can be positive or negative based on whether the rising transition T_{Sp} leads T_{Sn} or lags the latter. Typically for signed conversion, the flash-type TDC would require two chains of delay elements where one of the chains converts the positive time difference and the other converts the negative time difference. However, that leads to the use of a double amount of circuit components with twice the power consumption. Instead, the flash-type TDC's architecture is modified to contain only one chain of delays. Then the TDC converts the absolute time difference $|T_S| = |T_{Sp} - T_{Sn}|$ with the single chain of delays and uses a time-based comparator (D-flipflop) to find the polarity of input time difference whether it is positive or negative. Before digitization occurs, an arbiter, which is constructed with a combination of AND and OR gate, finds the leading edge between the rising transitions T_{Sp} and T_{Sn} and the lagging one. It generates two rising edges, T_{START} and T_{STOP}, where T_{START} always leads T_{STOP} and the time difference $(T_{STOP} - T_{START}) = |T_{Sp} - T_{Sn}|$. The delay elements ($\Delta T$) used in the TDC are implemented using two half-delay elements ($\Delta T/2$) in series, which are based on current-starved inverter cells as depicted in the bottom left corner of Fig. 6.11. The raw resolution of this course TDC, which is equal to the delay $\Delta T =\sim 6.25$ ns is adjusted by control voltage V_{cntrl}. The delays are subject to process and temperature-related variations and therefore calibrated using a DLL-based foreground technique. The TDC starts the conversion at the occurrence of T_{START}, and the chain of

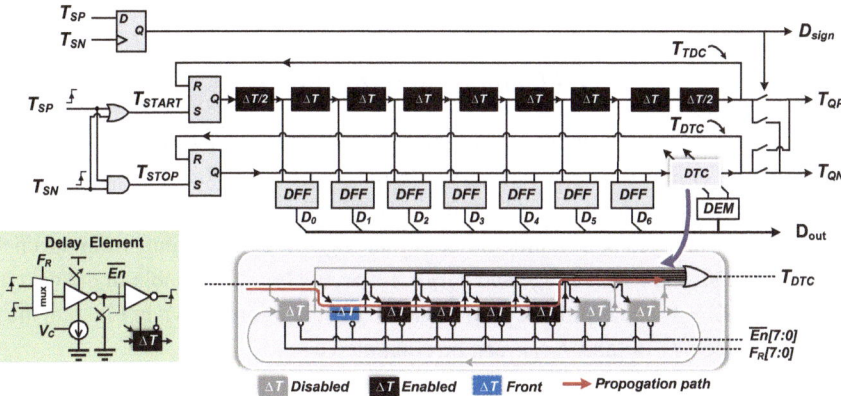

Fig. 6.11 Overview of Qe extractor circuit with the TDC and DTC

delays generates a series of rising edges, each following the previous one with a delay of ΔT. Only the first rising edge is generated after a delay of $\Delta T/2$. The conversion stops with the arrival of T_{STOP}, and it triggers seven D-flipflops to make time-domain comparisons and convert the time difference to its digital equivalent $D_{out}[6:0]$.

The ideal transfer characteristic of this flash-type TDC is illustrated in Fig. 6.12a. The TDC's input range is extended two times using D_{sign}, which identifies the polarity of T_S using a D-flipflop. Now, the rising-edge, T_{TDC} generated by the end of TDC's chain of delays represents a delayed version of T_{START} with a total delay amounting to $8 \times \Delta T$. The rising transition T_{STOP} also passes through a series of delay elements which is a part of the DTC driven by $D_{out}[6:0]$ and produces T_{DTC}. During operation, the total delay produced

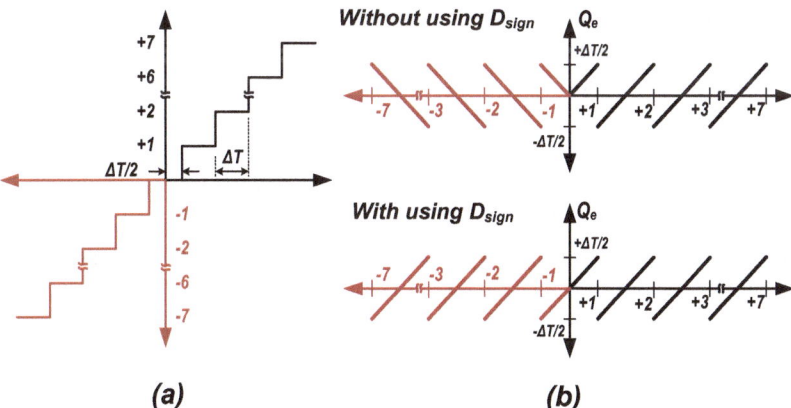

Fig. 6.12 a Ideal transfer characteristics of the TDC and **b** the resulting quantization error (Q_e) with and without using D_{sign}

by the DTC can be expressed as

$$DTC_{out}[n] = (8 - D_{out}[n])\Delta T$$

When the DTC's delay elements are disabled, they are bypassed, providing zero delays to the rising-edge T_{STOP}. The value of $D_{out}[6:0]$ ranges from 0 to 7. For $D_{out} = 0$, the DTC produces the maximum delay of $8\Delta T$, and for $D_{out} = 7$, the DTC produces the minimum delay of ΔT. The delayed pair of time-based signal edges, i.e., the outputs from the chain of delays in the TDC and DTC: T_{TDC} and T_{DTC}, respectively, are available in the next clock cycle when CLK $= 1$. Now, with the help of the timing diagrams shown in Fig. 6.8, the time difference between T_{TDC} and T_{DTC} can be computed as

$$T_{DTC}[n] - T_{TDC}[n]$$
$$= (T_{STOP}[n - 1/2] + (8 - D_{out}[n - 1/2])\Delta T) - (T_{START}[n - 1/2] + 8\Delta T)$$
$$= T_S[n - 1/2] - D_{out}[n - 1/2]\Delta T = Q_e[n - 1/2]$$

The signal edges T_{TDC} and T_{DTC} are interchanged according to the value of D_{sign} and generate two rising edges identified as T_{Qp} and T_{Qn}, respectively. For any clock-cycle "n", the time-difference $T_Q[n]$ between the rising edges $T_{Qp}[n]$ and $T_{Qn}[n]$ can be expressed as $-Q_e[n - 1/2]$. Now, the resulting quantization error, Q_e in combination with the value of D_{sign} is demonstrated in Fig. 6.12b.

6.3.3 Dynamic Element Matching Algorithm

The random mismatch among the delay elements in the TDC combines with the quantization noise but reduces due to high-pass noise shaping by the $\Delta\Sigma$ loop. Unlike the flash-type TDC, any mismatch among DTC delay elements produces distortions in the signal band and eventually reduces the maximum achievable dynamic range. A dynamic element matching (DEM) technique is employed on the DTC's chain of delay elements to minimize the mismatch-induced in-band distortions. Generally, the DEM technique, based on deterministic approaches like clock-level averaging (CLA) and data-weighted-averaging (DWA), modulates the mismatch components and up-converts those beyond the signal bandwidth so that the decimation filter can remove it. The alternative is the stochastic method, which randomly chooses elements using a pseudo-random sequence. In this method, the mismatch components are randomized and appear as in-band white noise. The CLA method starts with an index to select one of the elements, and the index is incremented each clock cycle. The modulated mismatch component in the frequency domain exists at f_s/N and its multiples, where N is the total number of elements. The CLA method is less complex to implement and, therefore, preferred over the DWA method. However, for multi-bit architectures with a much larger number of elements, the modulated mismatch components in the CLA method

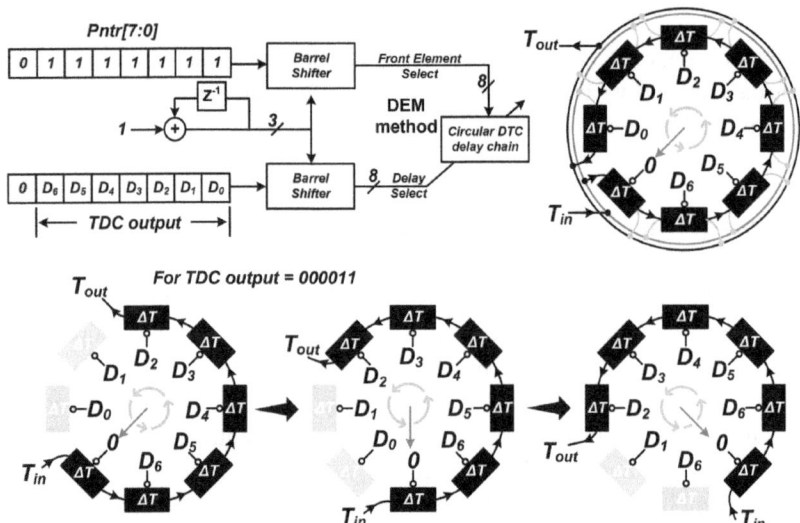

Fig. 6.13 An illustration of DEM technique used on DTC delay elements

fall within the signal bandwidth. There the DWA method is usually preferred regardless of its complexity in terms of implementation.

In this implementation, a CLA-based DEM technique is adopted to minimize the effects of mismatch in DTC. Figure 6.13 provides an overview of the DEM technique. As per the CLA technique, the delay elements inside the DTC are arranged in a circular queue where the pointer to the front element rotates anti-clockwise each clock cycle. The index of the pointer is generated with the help of a 3-bit ring oscillator driven by the global clock ($=$ 10 MHz) and a barrel shifter. Figure 6.14 provides a schematic representation of the barrel

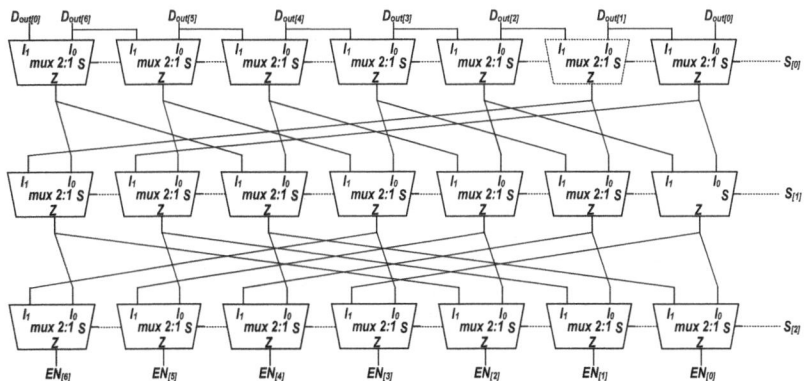

Fig. 6.14 A schematic of the mux-based barrel-shifter circuit

shifter. The circuit of the barrel shifter consists of three stages. The input data is rotated by
1 bit in the first stage, and in the second stage, it rotates by 2 bits. In the case of the third
stage, the data rotates by 4 bits. Overall, the barrel shifter can rotate the input bits of data by
a maximum of 7 bits as applicable in this implementation.

6.3.4 Foreground Calibration Circuits

The TDC and DTC delay elements (ΔT) are prone to process and temperature variations. A
DLL-based foreground calibration technique is employed to adjust and compensate for the
delays' value before a measurement phase. As illustrated in Fig. 6.15a, the first calibration
circuit consists of a chain of eight delay elements, a 16-bit thermometric encoded current
digital-to-analog converter (IDAC[15:0]), and a calibration control logic block. The calibration control logic contains a 16-bit shift register R[15:0] and a time-error detector circuit
built using a combination of two D-flipflops. As depicted in Fig. 6.15b, a reset triggers the
calibration phase and sets the output of the shift register output R[15:0] to 8. In the case

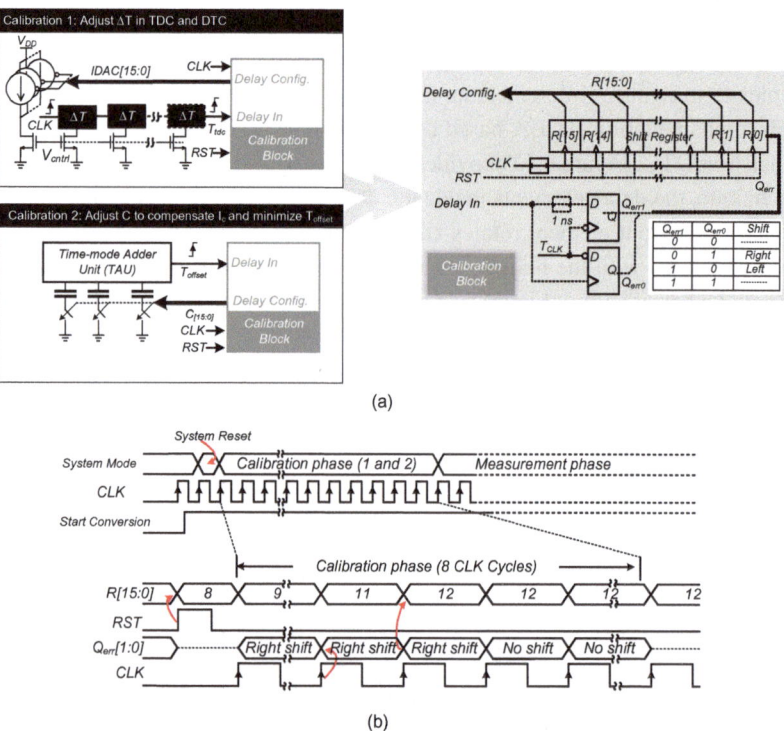

(a)

(b)

Fig. 6.15 a Architecture of the calibration circuits and **b** the timing diagrams during the calibration
phase

of ΔT calibration, the rising edge of the clock is passed through the chain of eight delay elements and then compared to the falling edge of the clock. If the total delay produced is less than $T_{CLK/2}$, the shift-register shifts right and decreases the value of IDAC[15:0] such that the control voltage to the delay is decreased and the total delay approaches the desired value. Initially, after resetting the shift register's output R[15:0] is set to 8, it continues another eight clock cycles to finish the calibration.

As mentioned earlier, T_{offset} produced at the output of a single-ended TAU circuit acts as a common mode time-based input for the next stage. Although T_{offset} is canceled out for the differential circuit, it reduces the range of the next stage's differential time-based input. Therefore, the 2nd DLL engages the capacitor-array (C[15:0]) used in the TAU elements to minimize T_{offset} within 0.5 ns and compensate for I_c variations. During the calibration, a rising edge is generated from a replica of a single-ended TAU element with zero inputs which is equivalent to T_{offset}. The transition edge is compared to the falling edge of the clock. Based on the comparison result of whether the time difference is more than 0.5 ns, the 16-bit shift register increases or decreases the value of the capacitor array C[15:0]. During measurement, both calibration phases precede the $\Delta\Sigma$-TDC conversion following a system reset and run parallelly. As depicted in Fig. 6.15b, both calibrations finish within eight clock cycles.

6.3.5   Differential VTC for On-chip Testing

During measurement, the differential time-based input to the proposed $\Delta\Sigma$-TDC is provided regarding the time difference between the rising edges of two PWM signals. The two PWM signals are generated from a differential relaxation-type VTC circuit which converts the input voltages to their time-based counterparts. A schematic representation of the VTC circuit is illustrated in Fig. 6.16a. The VTC follows three phases of operation identified as "clear", "sample", and "convert" as illustrated with the timing diagram provided in Fig. 6.16c. The $\Delta\Sigma$-TDC is designed to capture the input time difference during the ON-period of the clock and provide the digital output in the next clock cycle as it becomes high again. To comply with this operation, the differential VTC circuit samples the input differential voltages when CLK $= 0$ and converts them to equivalent pulse widths in the next cycle when CLK $= 1$. The VTC circuit's architecture is pseudo-differential, and each half contains a pair of VTC unit cells. At first, in the "clear" phase the capacitor C_0 is charged to V_{DD}. Next, in the "sample" phase, the input voltages (V_{in}) are sampled on C_0 (see Fig. 6.16a). During the "convert" phase, C_0 is discharged linearly using a constant current source till it crosses the switching threshold voltage (V_{TH}) of the next stage (NAND-gate). When the voltage crosses the threshold, a pulse width (T_{out})is generated at the output. The pulse width T_{out} is proportional to V_{in} and can be derived as

$$T_{out} = T_{CLK}/2 - \frac{V_{in} - V_{TH}}{\lambda}$$

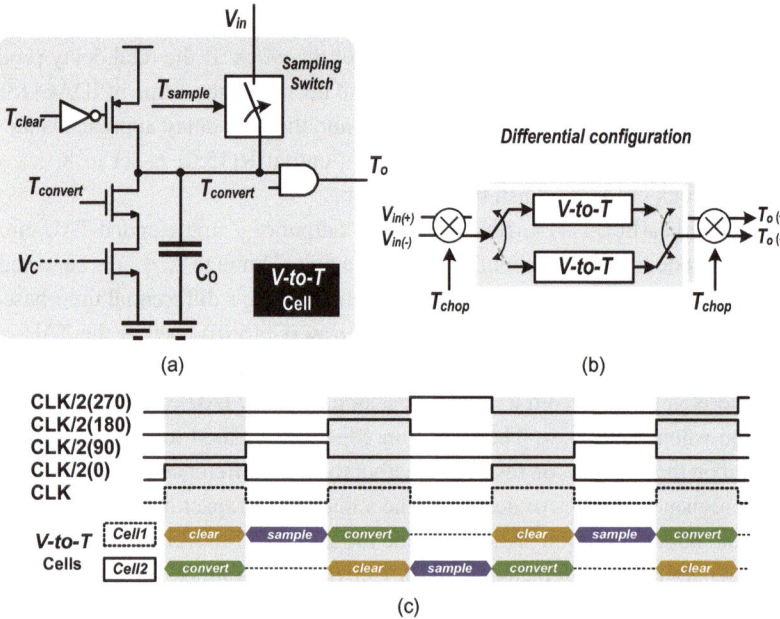

Fig. 6.16 A schematic representation of VTC with chopping and its timing diagrams

where λ is the discharging slew rate. In single-ended mode, a pair of VTC unit cells are interleaved to sample the input voltage alternatingly, and for differential mode, two pairs of VTCs are utilized. Now, the time difference produced by the differential VTC can be expressed as

$$T_{out}^+ - T_{out}^- = -\frac{1}{\lambda}(V_{in}^+ - V_{in}^-)$$

The conversion gain of the VTC circuit depends on the discharging slope λ. The voltage-domain noise translated to the time-based counterpart is also scaled with respect to λ. Therefore, a high slew rate was used in the implementation to reduce the input noise contribution. The differential VTC is also designed with "chopping" at the input and output terminals to remove mismatch-induced time-based offset errors.

6.4   Measurement Results

The proposed $\Delta\Sigma$-TDC prototypes were implemented in a 65 nm CMOS technology. The micrograph of the die (1×1 mm^2) with the enlarged TDC layout occupying an active area of 0.018 mm^2 is shown in Fig. 6.17a. Figure 6.17b shows the details of the test setup used during measurement. During measurement, the externally supplied 20 MHz ($2 \times CLK_{in}$) was divided on-chip to generate a 50% duty-cycled sampling clock of 10 MHz. The core

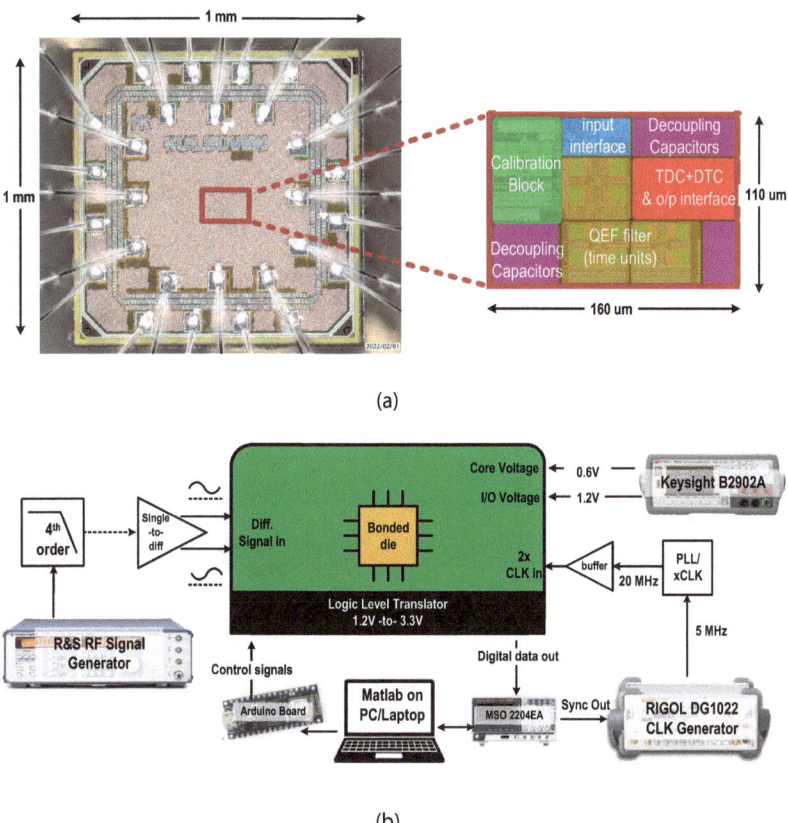

(a)

(b)

Fig. 6.17 a Micrograph of the die with an enlarged view of $\Delta\Sigma$-TDC layout and **b** the details of the measurement setup

of the $\Delta\Sigma$-TDC and the input-output (IO) interface were powered from 0.6 V and 1.2 V, respectively, generated from two channel precision source units (PSU) Keysight-B2920A. The digital bit-streams from the chip were level-shifted from 1.2 V to 3.3 V, suitable for recording by the logic analyzer "MSO 2204EA". The digital bit-streams of the 4-bit signed output from the TDC were further analyzed using "MATLAB" software on a PC or laptop. The digital configuration inputs and control signals to the bonded samples were provided using an "Arduino Nano 33-ToT" interface board.

Figure 6.18a, b, c provides the three output spectrums of the $\Delta\Sigma$ modulator before decimation for different orders of noise-shaping: first, second, and third orders, respectively. The spectral plots are generated from 2^{13}-pt FFT sequence using the Hanning window of the same length on the recorded samples of 2 ms. As shown in Fig. 6.18, the output spectrums in the case of (a) first, (b) second, and (c) third orders follow trends of 20 dB/dec, 40 dB/dec, and 60 dB/dec up-conversions, respectively. A 153 kHz differential 1.6 ns

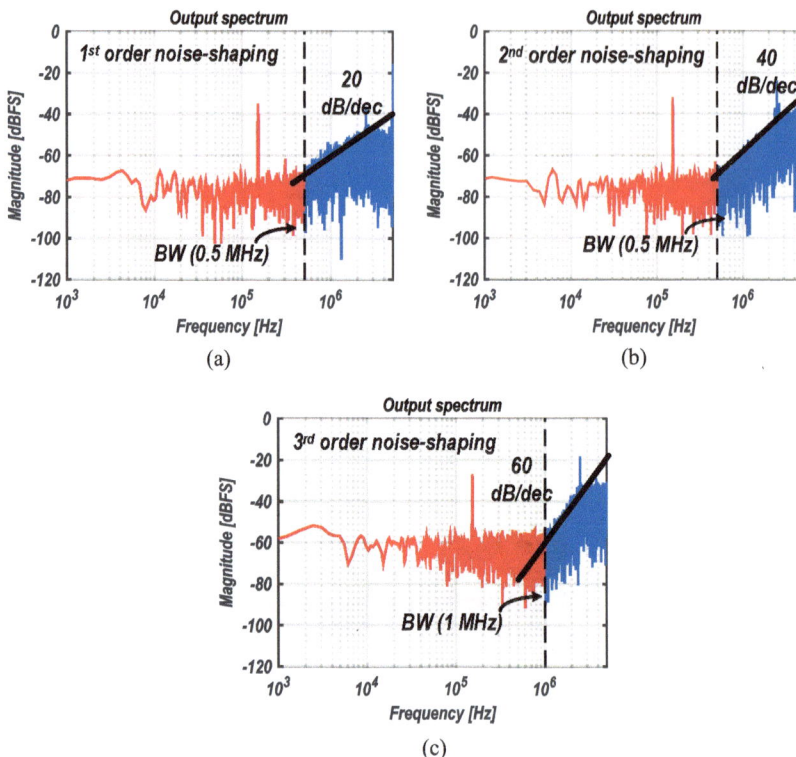

Fig. 6.18 Output spectrum of the $\Delta\Sigma$-TDC in the case of **a** 1st-, **b** 2nd-, and **c** 3rd-order noise-shaping

p-p sinusoidal time difference was used to measure the integrated noise in the presence of an input. The sinusoidally varying time difference was generated using the on-chip VTC from the externally supplied 10 mV p-p differential voltage over 0.9 V DC voltages. The integrated noise from the spectral plots as measured up to $f_{BW1,2} = 0.5$ MHz for first and second orders of the $\Delta\Sigma$ modulator is 114 ps_{rms} and 92 ps_{rms}, respectively. In the case of third order, it equals 418 ps_{rms} integrated up to $f_{BW3} = 1$ MHz. For third order, the device-noise contribution from additional time-based adders over $2 \times f_{BW1,2} (= 1$ MHz$)$ results in a larger in-band noise contribution.

The outputs from the $\Delta\Sigma$-TDC sample for the sinusoidal inputs (153 kHz differential 1.6 ns p-p time difference) are filtered by passing through a 500 kHz bandwidth low-pass filter. They are provided in Fig. 6.19a. The core of the $\Delta\Sigma$-TDC consumes 11 μW, 13.2 μW, and 15.8 μW power from 0.6 V supply while operating as first-, second-, and third-order $\Delta\Sigma$ modulators. The percentages of the power breakdown for various circuits for all three orders of noise shaping are illustrated in Fig. 6.19b. For the first order, the power consumed by Q_e extractor circuit is dominant. However, for higher orders of noise shaping, as additional

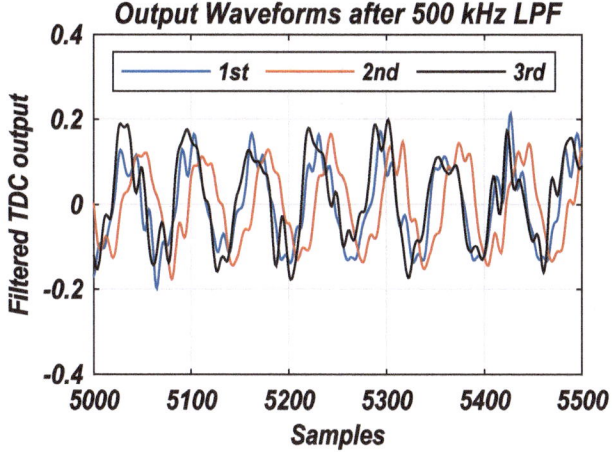

(a)

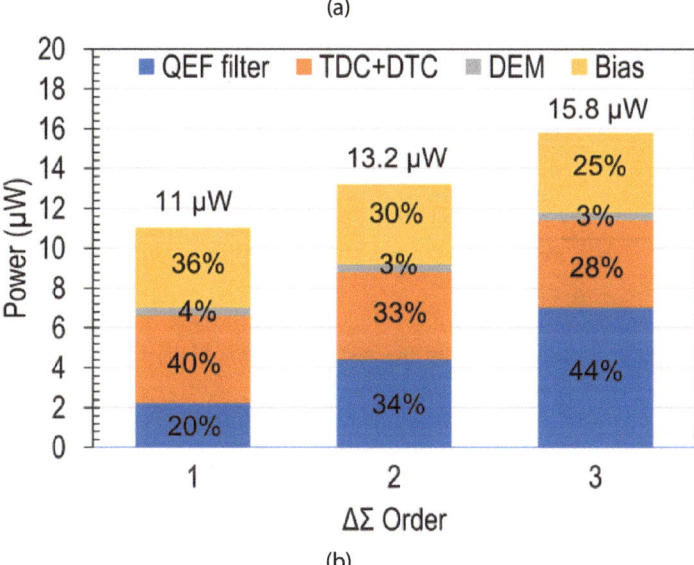

(b)

Fig. 6.19 a Filtered $\Delta\Sigma$-TDC output waveforms for different modes (first, second, and third orders) from a 153 kHz differential 1.6 ns p-p sinusoidal time-difference after passing through a 500 kHz bandwidth low-pass filter and **b** percentages of power breakdown for different circuit blocks

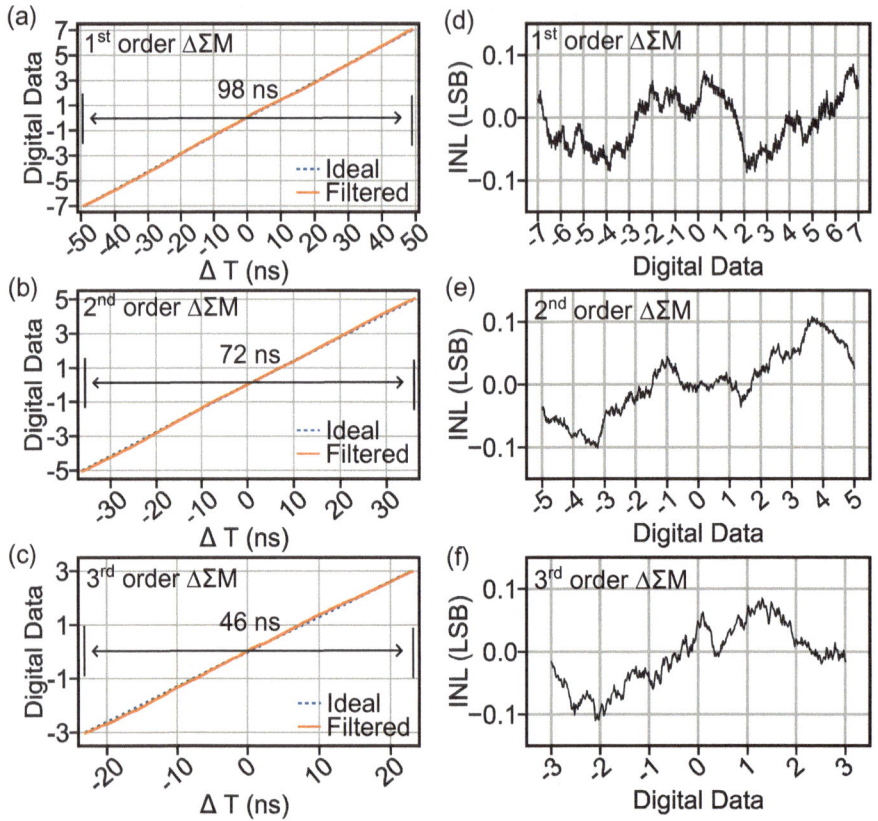

Fig. 6.20 $\Delta\Sigma$-TDC's Transfer characteristics **a–c** and measured INL data **d–f** of the TDC for different orders of $\Delta\Sigma$ modulation

time-based adders are made functional, the dominant unit dissipating the maximum portion of total power is the QEF filter.

The averaged digital output from the $\Delta\Sigma$-TDC and the INL data tested with ramp time-difference inputs for first, second, and third orders of $\Delta\Sigma$ modulation is illustrated in Fig. 6.20. The maximum measured full-scale T_{range} is 98 ($= \pm 49$) ns for the first-order $\Delta\Sigma$ modulator. During higher order operation of second and third orders, the input ranges of the $\Delta\Sigma$-TDC are chosen to be limited to 75% and 50% of the full-scale value $\pm T_{CLK}/2$ ($= \pm 50$ ns), respectively, to prevent saturation of the time-based adder units during discrete-time operation. The measured T_{range} for second and third orders of noise-shaping operation are 72 ($= \pm 36$) ns and 46 ($= \pm 23$) ns, respectively. The measured integral non-linearity INL for three different modes is within ± 0.1 LSB.

Performance Comparison with State-of-the-Art Designs

The performance metrics of the $\Delta\Sigma$-TDC are benchmarked with respect to the state-of-the-art $\Delta\Sigma$-TDCs with noise-filter orders ≤ 3 in Table 6.1. Compared to recent multi-bit state-of-the-art designs, this $\Delta\Sigma$-TDC with the least core area (0.018 mm^2) consumes power with a range of 11–15.8 μW (42x less in the worst case). The dynamic range (DR) in the case of time-based noise-shaping TDCs is computed based on the integrated rms noise over the signal bandwidth and can be expressed as

$$DR(dB) = 20log_{10}\frac{\text{Full-scale } T_{range}}{\text{Integrated rms noise}}$$

The measured DR for first, second, and third orders of $\Delta\Sigma$ modulation is 58.7 dB, 57.9 dB, and 40.7 dB, respectively. Note that similar to the approach followed in [91, 158, 159] $FOM_{2\pi}$ based on DR using the integrated rms noise, appears to be a better indication of performance metrics applicable for time-based architectures of $\Delta\Sigma$-TDCs. The figure-of-merit $FOM_{2\pi}$ as used in [91, 158, 159] can be expressed as

$$FOM_{2\pi} = \frac{Power}{min(f_s, 2f_{BW}) \times 2^{(DR-1.76/6.02)}}$$

Table 6.1 Performance comparison with state-of-the-art $\Delta\Sigma$-TDCs

	ISSCC' 12 [157]	TCASI' 14 [158]	JSSC' 18 [159]	TCASI' 20 [91]			This work		
TDC architecture	SRO	Gated SRO	T-to-V + CT	Flash-TDC + MASH-$\Delta\Sigma$			$\Delta\Sigma$		
Differential architecture	Yes	No	No	No			Yes		
CMOS process (nm)	90	65	65	40			65		
Sampling freq. f_s (MHz)	50	400	250	50			10		
Bandwidth f_{BW} (MHz)	1	4	1	2.5			0.5		1
Noise-filter order	1	2	3	1	2	3	1	2	3
Dynamic range DR (dB)	84	73	80.8	59.8	65.8	66.8	58.7	57.9	40.7
Full-scale range (ns)	5	0.666	2	0.32			98	72	46
Power supply (V)	1	1	1.2	1.1			0.6		
Core power (μW)	2000	6720	8400	670[i]	1000	1320	11	13.2	15.8[i]
TDC core area (mm^2)	0.02	0.05	0.055	0.04	0.06	0.08	0.018		
$FOM_{2\pi}$ (fJ/conversion-step)	77[ii]	228	470	170	130[iii]	150[iv]	15.6[ii]	20.6[iii]	89.2[iv]

Note: (i) Worst case ~42x less; (ii)~5x improvement; (iii)~6.3x improvement; (iv)~1.7x improvement

Despite limited DR (dB), this work achieves the best $FOM_{2\pi}$ of 15.6, 20.6, and 89.2 fJ/conversion-step for first, second, and third orders, respectively, which translate to ∼5x, ∼6.3x, and ∼1.7x improvements in the respective orders with reference to current state-of-the-art $\Delta\Sigma$-TDCs.

6.5   Conclusion

This work has proposed a differential noise-shaping TDC, implemented in 65 nm CMOS technology, and the prototype samples demonstrate a single-loop multi-order (≤3) $\Delta\Sigma$ modulation. The TDC core occupies an area of 0.018 mm^2. The measured maximum T_{range} of the $\Delta\Sigma$-TDC prototype is 98 (= ±49) ns. The proposed architecture uses time-based arithmetic units and is based on a time-based FIR filter. It is used in feedback to filter the quantization error resulting from digitization. The filter coefficients are adjustable and can be configured to realize first, second, and third orders of $\Delta\Sigma$ modulation. The TDC consumes power within the range of 11–15.8 μW from a 0.6 V supply at 10 MHz clock frequency for multiple orders (≤3) of $\Delta\Sigma$ modulation. The area efficiency and power are traded with increased noise levels in the signal band. In future, a trade-off must be explored to optimize the in-band noise within the limited power budget while maximizing the energy efficiency. Despite the limited DR (dB), to the authors' knowledge, this work achieves the best state-of-the-art energy efficiency with the range of 15.6–89.2 fJ/conversion step for first, second, and third orders, respectively, with reference to current $\Delta\Sigma$-TDCs having noise-shaping filters of the order ≤3. The proposed $\Delta\Sigma$-TDC samples are suited for integrating ultra-low-power area-efficient ADPLL designs particularly operating with sub-1V supplies.

Abstract

This chapter is focused on the design and development of a radiation-hardened, wide-band, low phase-noise fractional-N all-digital phase-locked-loop (ADPLL)-based frequency synthesizer for clock and pulse-width-modulated (PWM) signal generation. The prototype of the frequency synthesizer is implemented using a commercial 65 nm CMOS technology. The synthesizer features LVDS, LVCMOS, and PWM outputs and is targeted for use in various application areas as in space, nuclear power plants, and HEP experiments, and therefore hardened with respect to both TID and SEEs (SELs, SEUs, and SETs). The project is planned, designed, and implemented as a part of an industrial collaboration between the ADVISE research group of KU Leuven and MAGICS Technologies and funded by the European Space Agency (ESA). The fully integrated ADPLL comprises a Digitally Controlled Oscillator (DCO), Multi-Modulus Divider (MMD), Time-to-Digital Converter (TDC), digital loop filter, output dividers, buffers along with several bias generating circuits, and digital communication interfaces. I contributed to this project by designing the DCO, MMD, output dividers, and output buffers.

7.1 Introduction

Traditionally, PLLs are used to generate frequencies with high spectral purity, precision, and stability and play an essential role in building CDR circuits. The requirements for design specifications and parameters are much more stringent for highly reliable applications like in HEP experiments as well as communication circuits in space [161]. It requires additional

---

Parts of this chapter were adapted from the paper published in RADECS 2021 conference proceeding [160].

© The Author(s), under exclusive license to Springer Nature Switzerland AG 2024   105
A. Karmakar et al., *Integrated Time-Based Signal Processing Circuits for Harsh Radiation Environments*, Synthesis Lectures on Engineering, Science, and Technology,
https://doi.org/10.1007/978-3-031-40620-1_7

design efforts to provide increased immunity of performance metrics against radiation-induced effects (TID, SEE, etc.) while keeping the performance deviations under acceptable limits.

Conventional analog charge-pump-based PLL architectures are sensitive to ionizing radiation, both regarding SEE and TID effects [144, 162]. The analog phase detector's output node, in particular, is found to be sensitive to SEEs, which are likely to generate sudden phase transients. However, such fault conditions are recovered in due course by the closed-loop action of the PLL architecture. Moreover, the continuous exposure to TID leads to output frequency deviations which are eventually compensated by the negative feedback of PLL up to a limit when it exceeds the lock range. These radiation concerns for reliable operation must be dealt with additional care by incorporating different RHBD techniques.

As found in studies [17], LC-tank-based oscillators are much more immune to radiation-induced frequency deviations and therefore preferred compared to delay-based ring oscillators. Although ring oscillators could provide a wide tuning range that favors the implementation of wide-range frequency synthesizers and requires much less area, LC oscillators are still preferred considering their better phase noise performance and high spectral purity than their ring oscillator-based counterparts. Here, in this implementation, the wide-tuning range is one of the critical requirements. Therefore, the architecture of the LC oscillator is extended with a dual-tank-based implementation to accommodate large frequency tuning ranges. In addition, the MOS-varactors are avoided in the LC tank implementation. Instead, digitally switchable capacitor banks are used, which is a better implementation regarding radiation tolerance. Compared to a completely analog implementation of PLLs, all-digital-based architectures provide better performance reliability in radiation environments. In advanced technology nodes, the design and hardening of analog component blocks are challenging. The sensitive parts are replaced with their digital counterparts in ADPLL-based frequency synthesizers. These take advantage of the availability of systematic SEE hardening techniques for digital circuits, such as TMR and temporal redundancy [163], while continuing to exploit the performance gain provided by these nodes. In addition, integrating several digital signal processing (DSP) algorithms with digitally controllable performance calibration units becomes an effective tool to mitigate TID-induced degradation of functionalities.

The achievable frequency resolution is always one of the concerns for ADPLL-based implementations compared to their analog counterparts. Therefore, custom-designed fine capacitor banks are used [164], and a fractional-N-based closed-loop architecture is adopted in this design to facilitate finer frequency tuning. In addition, the spurs generated from the fractional modulation of the oscillator output frequency are minimized by introducing a $\Delta\Sigma$ dithering in the control signals of the frequency divider.

The architecture, design, and implementation of the individual blocks are explained in Sect. 7.2. The performance metrics and the results of radiation assessment of the designed prototype are discussed in Sect. 7.4, with concluding remarks in Sect. 7.5.

7.2   Architecture of the Frequency Synthesizer

The frequency synthesizer can generate a wide range of frequencies from a reference clock to support applications such as onboard clock generation, RF frequency synthesis, and clock jitter filtering. The circuit is fully integrated without needing any external loop filter or VCO, except for a crystal with loading capacitors and power supply decoupling components. Figure 7.1 shows the overview of the architecture of the radiation-hardened ADPLL. It works based on the principle of fractional-N PLL-based frequency generation. It comprises the major building blocks such as (i) a reference clock frequency generator, (ii) an ADPLL with on-chip DCO, (iii) output clock dividers and drivers, (iv) a digital PWM clock generator, (v) a serial communication interface, and (vi) a digital core to control, debug, and calibration. As the PLL interface requires, the reference frequency to sync the free-running DCO inside the ADPLL structure is generated from a low-jitter temperature-compensated clock source. The on-chip option can generate a reference clock by connecting an external crystal (48 MHz) to the chip. Alternatively, the reference frequency can be generated directly from an externally interfaced differential CMOS-compatible clock signal (20 MHz–66 MHz). Various single-ended and differential clock signals (CMOS, LVDS, CML, etc.) are supported with selectable voltage ranges of 1.2, 1.8, 2.5, or 3.3 V.

The reference clock signal is connected to one of the inputs of the TDC placed inside the closed loop ADPLL core. The ADPLL comprises a TDC, a digital loop filter, a dual-core DCO in the forward path, an AC-coupled CMOS buffer, a fixed divide-by-2 block, and a programmable MMD in the feedback path. The MMD takes in the buffered and divided oscil-

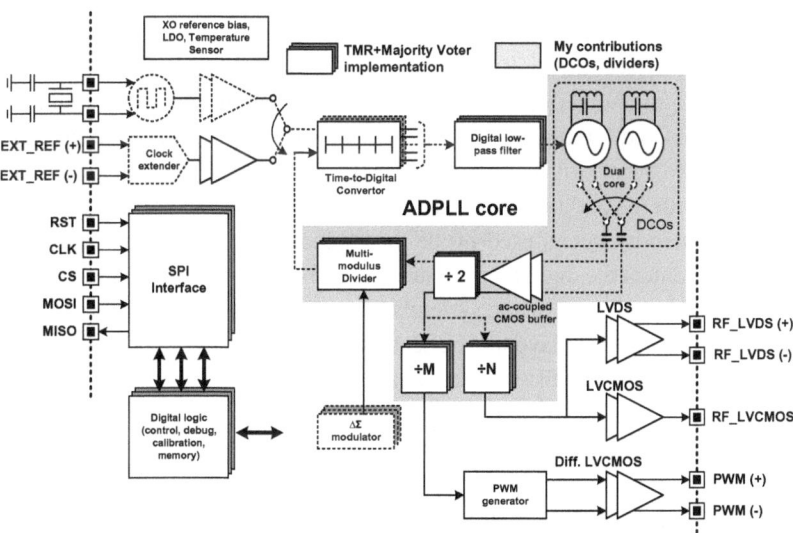

Fig. 7.1 Architecture of the radiation-hardened fractional-N ADPLL

lator output and generates feedback regarding the low-frequency clock signal to the TDC for phase-error measurement. The programmable modulus divider is driven and configured by a $\Delta\Sigma$ modulator to reduce the reference spurs around the clock frequencies in the frequency spectrum. The TDC measures the phase difference between the reference clock and the on-chip oscillator-generated frequency. In this design, the TDC architecture is implemented based on the principle of first-order noise-shaped time-based quantization using gated-ring-oscillator (GRO) [165]. The TDC aims to achieve a sub-picosecond resolution with oversampling, feasible for targeted applications in industrial and space environments. The TDC output is fed to the digital-domain loop filter, which drives the control signals of the DCO. The frequency synthesizer also features several on-chip output drivers: LVCMOS and LVDS. The LVCMOS output buffer is powered by a 1.2 V power supply and can deliver a single-ended output clock with a frequency ranging from 1 MHz to 200 MHz. A 2.65 V power supply powers the LVDS output driver, and it can produce clock signals with a frequency ranging from 1 MHz to 2.5 GHz. Its range can be extended further by bypassing the internal divide-by-2 block, leading to increased phase noise contributions. The PWM output is generated from the DCO, and the pulse width is configured using an 8-bit digital control word D_{PWM}. To generate the PWM signal, the output of an 8-bit counter is compared with D_{PWM}, and the output of the signal is toggled as soon as it crosses the value of the control word. As illustrated in Fig. 7.1, the chip also includes a serial communication (SPI) interface and a digital core to read and write the various configuration registers and thus be able to control, debug, and calibrate different interfaces.

The frequency synthesizer's various building blocks are radiation-hardened to minimize performance variation and functional failures over the exposure time. An LC-tank-based oscillator is chosen for the ADPLL core, as such architectures have been reported in literature [17, 134] to showcase minimal variation when exposed to high TID levels. The SPI communication interface, the digital core, and the digital loop filter are implemented with TMR. The TDC, programmable MMD unit, and the output dividers follow similar implementation patterns and interfaced with the majority of voters to filter out any soft errors arising from single event strikes. The use of TMR increases the dynamic power consumption three-fold with the addition of the dynamic power of the majority voters. Therefore, the number of majority voters used inside these blocks is minimized. Wider devices are also used to implement the digital logic gates inside these blocks to minimize TID-induced parameter variation and elongate the system's lifetime and endurance against high TID levels. While implementing the layouts, the sensitive regions are designed with double guard rings ($n + / p+$) to prevent SEL-induced failures.

7.3   Circuit Implementation

7.3.1   Digitally Controlled Oscillators (DCOs)

The architecture of the DCO implementation (Fig. 7.2) inside the ADPLL core is based on a class AB-type LC oscillator with NMOS-PMOS cross-coupled pair and an NMOS current source [166]. The frequency of the DCO can be digitally tuned with discrete steps utilizing switchable capacitor banks in the LC tank. Two different DCOs are implemented with overlapping tunable frequency bands. The frequency of the lower band DCO (DCO_LB) ranges from 2.5 GHz to 3.8 GHz, and the higher band DCO (DCO_HB) is tunable from 3.6 GHz to 5 GHz, and together they are tunable from 2.5 GHz to 5 GHz. The DCOs comprise three capacitor banks: (i) a binary-weighted coarse bank, (ii) a unary-weighted medium bank, and (iii) a unary-weighted fine bank. An 8-bit control word is used to configure the course bank with a maximum frequency resolution of approximately 20 MHz. The medium bank tracks the frequency deviation due to temperature variation and covers at least three LSBs (> 60 MHz) from the previous bank and the maximum frequency resolution of approximately 1 MHz. The fine bank is used to acquire the frequency with a much finer resolution of 50 kHz and is expected to cover a 3-LSB frequency range (3 MHz) from the previous bank. The medium and fine capacitor banks are configured using a 64-bit thermometer-coded control word. The digital frequency control inputs to the DCOs are driven by a second-order $\Delta\Sigma$ modulator to achieve a finer frequency resolution (< 50 kHz) by time averaging and quantization noise shaping of the capacitors in the fine bank.

As required in this implementation, the duty cycle of the oscillator's outputs must be adjusted to 50 ± 1 %. As illustrated in Fig. 7.3, the clock signal generated from the oscillator is AC-coupled to standard two-transistor-based inverter inputs. The PMOS and NMOS devices of the inverter are sized accordingly to be of equal strength. Then the resistor (R) is placed in the feedback to set the DC operating point of the transistors at $V_{gs} = V_{dd}/2 = 0.6$ V. The inverter with R-C feedback acts as an active high pass filter. The value of the capacitor (C) is chosen such that $f_{cutoff} < 1$ GHz to be able to pass the DCO frequencies. Two cascaded stages of AC-coupled inverters set the DC-value of the output clock signal at $V_{dd}/2$ and thus maintain the duty cycle 50 ± 1 %. While in operation, either of the DCOs is powered on based on the frequency requirement, and a tri-stated inverter is used to select the relevant output clock frequency.

The overview of the layouts of both DCOs is shown in Fig. 7.4. The core of the DCOs is composed of cross-coupled NMOS-PMOS pairs, which are placed near the inductor terminals. Arrays of decoupling capacitors surround each DCO unit. The layout of the course and medium capacitor banks of both the DCOs are implemented with metal-oxide-metal(MOM)-type capacitors. While designing the layout, the cells of the course capacitor banks are divided into five binary-weighted LSB cells and three thermometric MSB cells to minimize mismatch-induced variations. With this arrangement, the binary-weighted LSB cell of the course capacitor bank achieves a capacitance difference (ΔC_{CB}) of 13 fF, which

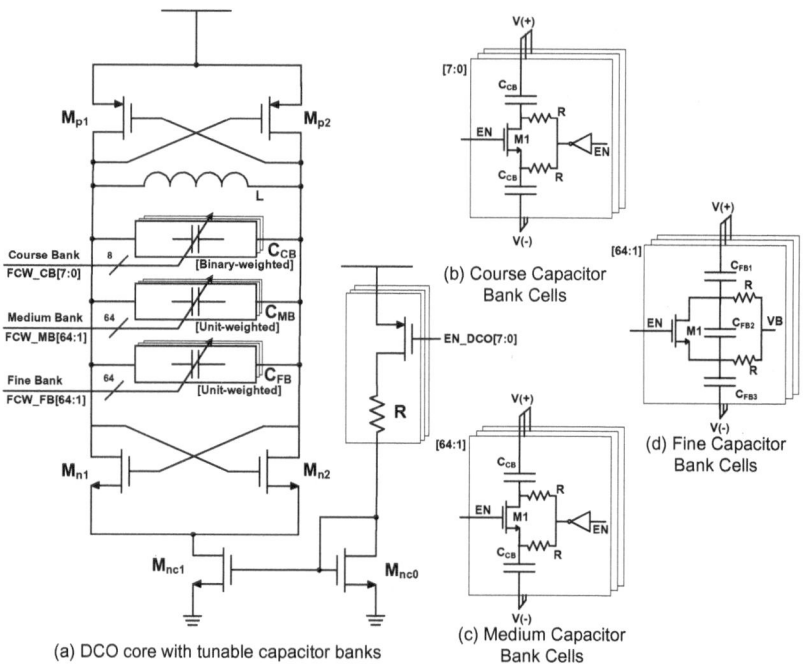

Fig. 7.2 Schematic of the **a** DCO cross-coupled core with various tunable capacitor bank cells: **b** course capacitor bank, **c** medium capacitor bank, and **d** fine capacitor bank

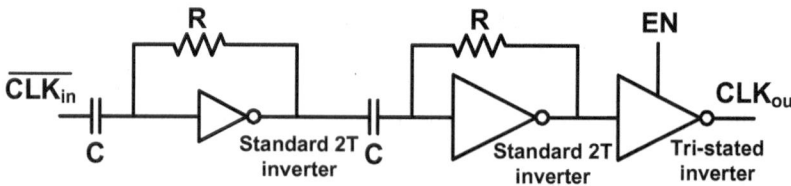

Fig. 7.3 Schematic of the AC-coupled CMOS buffer with tri-stated outputs

results in a frequency step of 13 MHz for DCO_LB and 10.6 MHz for DCO_HB. The medium capacitor banks are implemented with minimum standard MOM-type capacitors available within the technology database, resulting in a capacitance difference (ΔC_{MB}) of 1 fF, which causes a frequency step change of 1 MHz for both the DCOs. The layout of the unit cells of the fine capacitor banks is also MOM-type, but custom-made as used in [163] based on the fringe capacitance formed between narrow metal strips. Each cell consists of four metal strips (minimum width) formed in parallel with each other with a grounded poly-silicon layer to prevent noise coupling through the substrates. The outer two metal strips are connected to the oscillator nodes, whereas the inner ones are connected to an NMOS device. With this implementation, a switchable capacitance difference (ΔC_{FB}) of

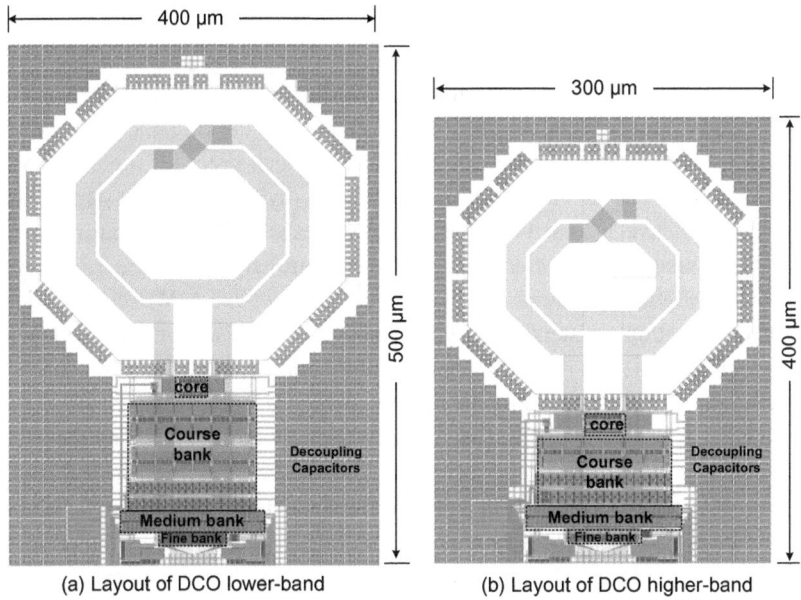

(a) Layout of DCO lower-band   (b) Layout of DCO higher-band

Fig. 7.4 Overview of the layouts of the DCOs (DCO_LB, DCO_HB)

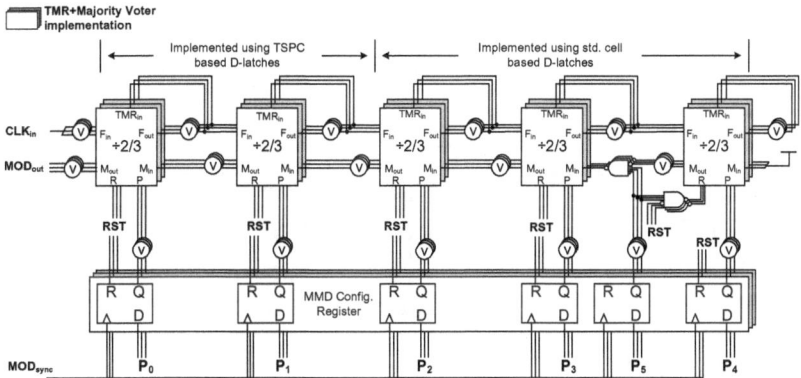

Fig. 7.5 Schematic of the triplicated MMD with TMR majority voters

0.06 fF capacitance is achieved while enabling or disabling thermometric unit cells of the fine capacitor. These cells achieve a fine frequency resolution of 62.5 kHz for DCO_LB and 100 kHz for DCO_HB. The mismatch generated variation among various cells of the capacitor banks is minimized by a closely matched layout with dummy units.

7.3.2 Multi-modulus Divider (MMD)

As required inside the closed loop architecture of the ADPLL, an MMD[167] is used to divide the output frequency of the DCO to the vicinity of the reference frequency fed to the TDC. As shown in Fig. 7.1 a divide-by-2 circuit block precedes the MMD unit. The implemented MMD circuit block is a programmable divider with an extended structure [168] consisting of five successive divide-by-2/3 stages with a division ratio ranging from 16 to 63. The division ratio is controlled by 6-bit control signals ($P_5 P_4 ... P_0$) driven by a $\Delta\Sigma$ modulator. Figure 7.5 shows the overall schematic of the MMD circuit with the 6-bit MMD configuration registers. Each MMD unit cell comprises four D-type latches with AND-logic gates connected in a closed loop configuration as shown in Fig. 7.6a. F_{in} from the previous stage triggers and starts the division, while M_{in} from the next stage is used to end the division. It can be configured as divide-by-2 or divide-by-3 logic by the $P_{[5:0]}$ input signal.

The first two stages of the MMD units are implemented with true-single-phase clock TSPC logic-based D-type latches as illustrated in Fig. 7.6b. The rest of the divider chain, i.e., three stages, are designed using transmission-gate-based conventional D-type latches as shown in Fig. 7.6c. Compared to transmission-gate-based structures, TSPC-based latches are driven by a single clock phase and have fewer MOSFETs. TSPC-based latches generate

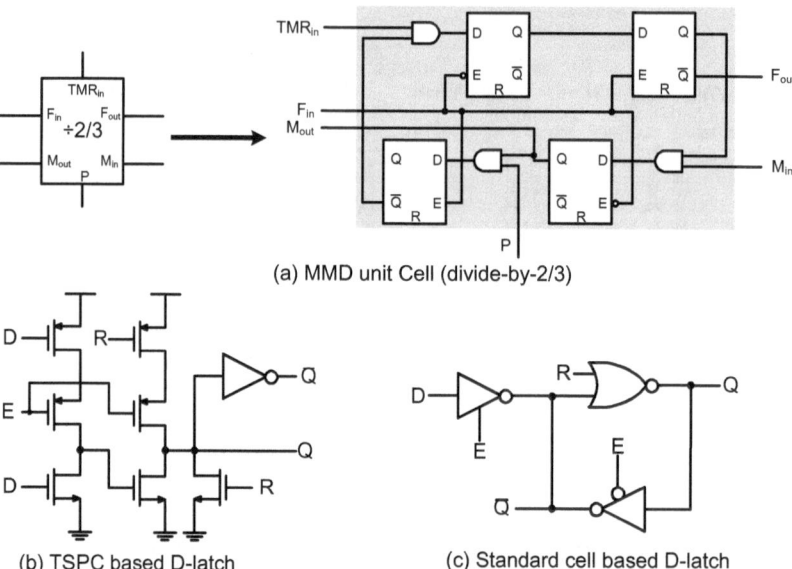

(a) MMD unit Cell (divide-by-2/3)

(b) TSPC based D-latch   (c) Standard cell based D-latch

Fig. 7.6 Schematic of the MMD unit cell implemented with TSPC logic gates and standard cells

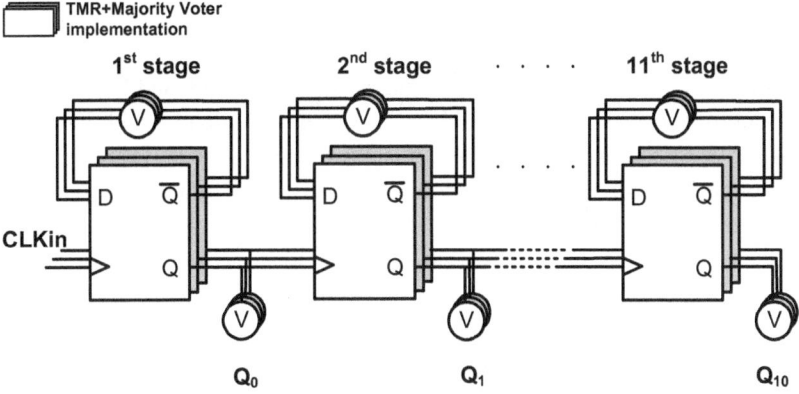

Fig. 7.7 Schematic of the 11-bit output divider with TMR majority voters

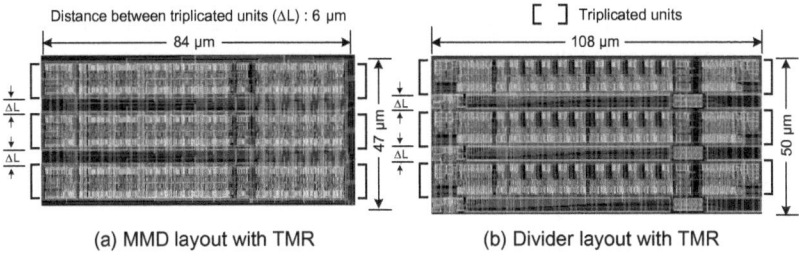

(a) MMD layout with TMR   (b) Divider layout with TMR

Fig. 7.8 Overview of the layouts of the triplicated MMD and output divider with TMR majority voters

much less propagation delay and are suitable for achieving the division operation at high frequencies (> 500 MHz). However, TSPC-based structures suffer from the data-loss problem while operating in low frequencies (< 500 MHz). It happens due to leakage-current-induced discharges at internal nodes between two consecutive clock edges and often leads to functional failures of the frequency divider. As shown in Fig. 7.5, each stage of the divider chain is triplicated and hardened against SEE-induced errors using TMR with the majority of voters placed in feedback. The number of majority voters used inside the divider chain is minimized to optimize the dynamic power dissipation. The overview of the layout (84×47 μm^2) of the MMD is presented in Fig. 7.8a. As illustrated in the implemented layout, each of the three layout units from the divider chain is separated by a distance of 6 μm to improve its tolerance from radiation-induced multi-cell upsets.

7.3.3 Output Divider

The integrated frequency synthesizer featuring LVDS, LVCMOS, and PWM drivers must deliver clock frequencies as low as 1 MHz. This requires a divider with a division ratio ranging from 2^0 to 2^{11}. In order to achieve this, the outputs of the DCO are passed through 11 stages of cascaded divide-by-2 binary dividers. An 11-to-1 multiplexer delivers one of these divided outputs to the output drivers and buffers. To comply with the requirements of high-frequency operation, the first two stages of the divider are implemented with TSPC logic-based D-flipflops. The subsequent nine stages of the divider are implemented using transmission-gate-based conventional D-flipflops considering the low-frequency operation. Each divider stage is hardened against SEE-induced errors using TMR, with the majority of voters placed in feedback.

Figure 7.7 shows the schematic of the 11-bit output divider comprising triplicated unit cells with majority voters for TMR implementation. The overview of the layout of the output divider is presented in Fig. 7.8b. As seen from the implemented layout, each of the three layout units of the cascaded divider chains is separated by a distance of 6 μm to reduce the sensitivity to multi-cell upsets.

7.4　Measurement Results

The ADPLL-based frequency synthesizer is implemented in a commercial 65 nm CMOS Technology. Figure 7.9a shows a die sample bonded inside the package and an enlarged layout of the ADPLL core (DCOs, TDC, MMD, and digital filter). A summary of the key performance metrics as obtained from the measurements is given in Table 7.1.

A 1.2 V supply powers the ADPLL core, whereas the IO interfaces use a 2.65 V power supply. During measurement, the frequency synthesizer is configured to generate a clock frequency of 2.5 GHz at the LVDS output. Then the phase noise performance is measured,

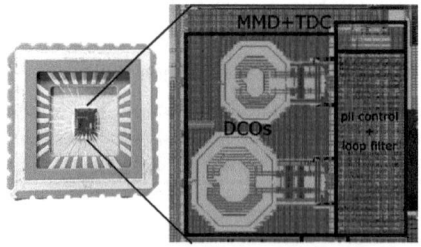

(a) Bonded die and enlarge view of the ADPLL layout

(b) Test setup for ADPLL samples at UCL facility

Fig. 7.9 Photograph of the bonded die sample inside its package and the radiation test setup for the ADPLL samples at the heavy ion facility at UCL

Table 7.1 Key performance metrics of the ADPLL

Metric		Results
Core/IO supply		1.2 V / 2.65 V
Frequency	LVDS	1 MHz–2.5 GHz
Range	LVCMOS	1 MHz–200 MHz
Phase noise at 2.5 GHz	@ 10 kHz	−101 dBc/Hz
	@ 100 kHz	−99 dBc/Hz
	@ 1MHz	−119 dBc/Hz
	@ 10 MHz	−131 dBc/Hz
Integrated Jitter [10 kHz–10 MHz]		460 fs
TID hardness		> 1 kGy
SEU/SEL hardness		>62.5 MeV.cm^2/mg

which equals −101 dBc/Hz at 10 kHz and −99 dBc/Hz at 100 kHz offset frequencies. An integrated jitter of 460 fs is measured within the loop bandwidth of 10 kHz to 10 MHz using the on-chip LC oscillator with the reference frequency generated from a 48 MHz crystal. The measured total power consumption of the chip including the IO interfaces is 90 mW. During the TID experiment, the phase noise performance of the closed loop ADPLL was measured at the 2.5 GHz oscillation frequency. The sample was subjected to a ^{60}Co gamma radiation source. The phase noise curves from the synthesizer's LVDS outputs were recorded before and after 1 kGy TID followed by accelerated aging phases. Figure 7.10a shows the phase noise characteristics. It can be seen that the variations between both distinct measurements are within 3 dB. The robustness of the frequency synthesizer samples against SEEs is evaluated using the heavy ion facility at the Université Catholique de Louvain (UCL). Two un-irradiated samples were subjected to a heavy-ion SEE assessment, powered at their nominal supply voltage of 1.2 V. Figure 7.9b shows the photograph of the mounted samples before removing the package lids and entering the vacuum chamber at UCL.

The on-chip SEE-hardened memory registers and serial communication interfaces are tested during the SEE experiment. The memory registers implemented within the chip are written with random bit patterns and validated by reading back the values. The experiment is done 10 times, and each time the samples are irradiated with an ion (LET = 62.5 MeV·cm^2/mg) flux of 10^3 ions/s/cm^2 for 3 seconds. No single-event-caused upsets are observed during this measurement window. The test for tolerance with respect to SEL was also performed. The heavy-ion testing was done with an ion flux of 10^4 ions/s/cm^2 for a measurement interval of 100 seconds till the fluence reaches the level of 10^6 ions/cm^2. The power supply currents are continuously sampled at each 5-second interval during the irradiation time at ambient temperature conditions. The different circuit blocks (e.g., digital communication interface, reference generation, synthesizer core, DCO, LVDS drivers, and CMOS drivers) were connected to different power supply channels to isolate and identify

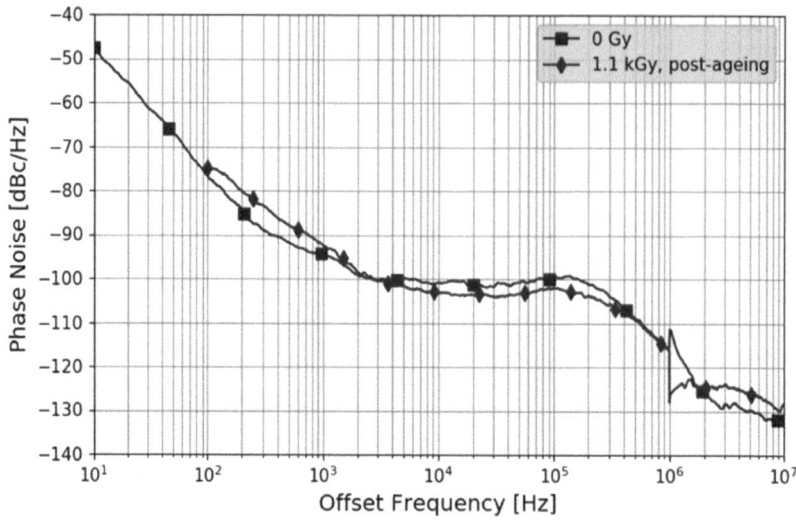

Fig. 7.10 Effects of TID on the phase noise of the ADPLL

SEL-sensitive chip areas. All the power supply current sources are monitored. Additionally, a protection limit was set on all supply currents, resulting in a power-down state if any of the readings crossed a specified limit. A multi-channel lab power supply with a sensing resolution of 100 μA is used during the testing period, and as the measurement results suggest, no increase in current readings is recorded. This is validated for all LET levels up to the maximum achievable from the facility, i.e., 62.5 MeV.cm^2/mg.

Performance Comparison with State-of-the-Art Designs

A performance comparison of the ADPLL circuit with respect to other state-of-the-art radiation tolerant PLL/CDR and ADPLL designs is provided in Table 7.2. The integrated oscillators are built using switchable dual-core-based architecture and have achieved the widest tuning range compared to their counterparts so far reported in the literature. The implemented DCO cores included in the ADPLL also achieve the best oscillator FOM. The implemented sample prototypes are intended for use in space-related applications, and therefore the TID evaluation is conducted till the qualification limit of 100 krad (= 1 kGy) is reached.

Table 7.2 Comparison with radiation tolerant PLL/CDR and ADPLL designs

Reference	[144]	[169]	[143]	[163]	This work
Topology	Analog CDR	Analog CDR	Analog PLL	ADPLL/ CDR	ADPLL
Oscillator	LC VCO	Ring VCO	LC VCO	LC DCO	LC DCO (dual-core)
Technology (nm)	65	65	65	65	65
Frequency range (GHz)	4.9–5.2	0.8–2	4.8–6	–	2.5–5
PN @ 1 MHz (dBc/Hz)	−122	–	–	−123.3	−119
Jitter (ps rms)	0.35	6.7	3.5	0.5	0.46
Power (mW)	34	7	18	11	8.5[†]
Oscillator FOM[§] (dBc/Hz)	−180	–	–	−183	−186
TID tolerance (Mrad)	350	600	250	1500	0.1
SEU/SEL tolerance (MeV·cm²/mg)	62.5	–	–	62.5	62.5

[†] post-layout estimation of the oscillators

[§] FOM = Phase Noise@$\Delta f - 20\log(\frac{f_o}{\Delta f}) + 10\log(\frac{Power}{1\text{mW}})$

7.5   Conclusion

An all-digital PLL-based fractional-N frequency synthesizer is implemented for application in harsh ionizing radiation environments such as space, nuclear facilities, and high-energy physics experiments. Different RHBD techniques at the transistor, topology, circuit, and layout levels are employed while designing the different functional units in the prototype. An assessment of up to 1 kGy of TID level has shown the robustness of the architecture in terms of frequency stability and phase-noise performance. The radiation tolerance of the samples is validated in terms of SEL and SEU sensitivity up to an LET level of 62.5 MeV·cm²/mg. The rad-hard frequency synthesizer chip can generate a wide range of LVDS output clock frequencies (1 MHz–2.5 GHz) to support applications such as onboard clock generation, RF frequency synthesis, clock cleaning, and jitter filtering. With 2.5 GHz at the LVDS output, the measured phase noise performance is −99 dBc/Hz at 100 kHz and −119 dBc/Hz at 1 MHz offset frequencies. An integrated phase jitter of 460 fs is measured using the on-chip oscillator. With a 48 MHz reference crystal, the overall power consumption of the chip, including the IO interfaces, is 90 mW.

Conclusion

8

Abstract

This book investigates and elaborates upon the research activities conducted during this study. The primary goals were to research and analyze the vulnerabilities of mixed-signal circuits (such as clock generators, data converters, etc.) concerning ionizing radiation and to develop new architectures utilizing various techniques which harden mixed-signal circuits against radiation. TID effects are the outcome of prolonged exposure to ionizing radiation. These effects cause a gradual decline in the performance of semiconductor devices. When a charged particle hits a sensitive circuit node, it can cause SEEs, which are brief disruptions produced by the accumulation of charge in semiconductor materials. In most circumstances, the resulting voltage or current transients are not destructive in nature. However, in a small number of instances, when the transients are strong, they can contribute to destructive phenomena like SEL and SEGR. The research was conducted with the end goal of finding practical applications in a variety of harsh radiation environments. Some examples of these environments include high-energy physics experiments, critical long-term space missions, and future nuclear-powered plants. The findings of the research, as discussed in this book, offer a comprehensive overview of the effects of radiation on various circuits and offer directions on how to make circuits more resistant to ionizing radiation.

This book investigates and elaborates upon the research activities conducted during this study. The primary goals were to research and analyze the vulnerabilities of mixed-signal circuits (such as clock generators, data converters, etc.) concerning ionizing radiation and to develop new architectures utilizing various techniques which harden mixed-signal circuits against radiation. TID effects are the outcome of prolonged exposure to ionizing radiation. These effects cause a gradual decline in the performance of semiconductor devices. When a

© The Author(s), under exclusive license to Springer Nature Switzerland AG 2024   119
A. Karmakar et al., *Integrated Time-Based Signal Processing Circuits for Harsh Radiation Environments*, Synthesis Lectures on Engineering, Science, and Technology, https://doi.org/10.1007/978-3-031-40620-1_8

charged particle hits a sensitive circuit node, it can cause SEEs, which are brief disruptions produced by the accumulation of charge in semiconductor materials. In most circumstances, the resulting voltage or current transients are not destructive in nature. However, in a small number of instances, when the transients are strong, they can contribute to destructive phenomena like SEL and SEGR. The research was conducted with the end goal of finding practical applications in a variety of harsh radiation environments. Some examples of these environments include high-energy physics experiments, critical long-term space missions, and future nuclear-powered plants. The findings of the research, as discussed in this book, offer a comprehensive overview of the effects of radiation on various circuits and offer directions on how to make circuits more resistant to ionizing radiation.

To be precise, the four major contributions of this research can be mentioned as follows:

- Chip I: Two different types of quadrature LC oscillators were implemented in 65 nm CMOS technology, and the sensitivity of different performance parameters with respect to TID was studied using X-ray irradiation. This study is the first comprehensive experimental investigation of TID-induced performance variations in LC-tank-based quadrature phase VCOs that has ever been reported in the published scientific literature [109].
- Chip II: A prototype of a radiation-hardened fractional-N frequency synthesizer for onboard clock generation was implemented as a part of an industrial collaboration with the financial support of ESA. This IC will likely be the first commercially available fully integrated radiation-hardened frequency synthesizer targeted for vital space missions, small satellites, and nuclear-related markets [160, 170].
- Chip III: A radiation-hardened time-based dual quantization $\Delta\Sigma$-CDC was proposed and designed in 65 nm CMOS technology suitable to be used during high-energy physics experiments. The prototypes were verified experimentally and achieved the fastest conversion time among the state-of-the-art designs and showed radiation tolerance in terms of SEL threshold up to 65 MeV·cm^2/mg [145, 146].
- Chip IV: A differential time-based $\Delta\Sigma$-TDC with configurable noise filter using time-based FIR feedback was proposed and implemented in 65 nm CMOS technology suitable for ultra-low-power applications. The prototypes were verified experimentally, occupied the least die area, and achieved the best figure-of-merit in terms of energy efficiency across different modes of operation.

The major takeaways from the outcome of the study and measurement results of developed IC prototypes are summarized in the next section. A discussion on research valorization and prospects for future works is also provided in the subsequent sections.

8.1   General Conclusion

This book provides a concise overview of the fundamentals of ionizing radiation (sources, effects on semiconductor devices, radiation sensors, and mitigation methods). In addition, it introduces to the readers the details of the time-based signal processing technique and elaborates on its applicability and effectiveness in the context of ionizing radiation mitigation. It provides several mitigation strategies and recommendations for reducing radiation-induced effects in ICs, with the implementation of radiation-hardened-by-design at the circuit level as the primary focus of the discussion. The chapter's content should make it easier for the readers to comprehend the various radiation-induced effects and recognize how critically important radiation hardness is. It also includes a coverage of numerous approaches and devices used in measuring radiation, primarily emphasizing semiconductor-based radiation sensors. The presented methods and recommendations are substantiated with experimental data from IC prototypes.

The readers are provided with some background information on the motivation behind the research study as well as an introduction to the subject of ionizing radiation in the introductory Chap. 1. The discussion begins by highlighting some of the most important landmarks and providing a high-level overview of the progression of ionizing radiation research over time. In addition to this, this chapter presents an overview of the work packages and activities as parts of the RADSAGA ITN initiative. The primary goals and objectives of the research program have also been discussed at length there. A 65 nm CMOS technology has been selected for the design and development of the circuit, as described in Chap. 1. Additionally, numerous time-domain signal-processing techniques have been explored for circuit-level implementations. Considering the radiation hardness requirement of the prototype designs, this chapter also elaborates upon the rationale behind the choice of specific CMOS technology and signal-processing technique.

Chapter 2 starts with briefly explaining the mechanisms of interaction of radiation particles with semiconductor devices. This chapter also discusses the sources of ionizing radiation and provides an overview of the various radiation-related effects (TID, SEE, and DD) that can affect electronic circuits. The chapter's content should make it easier for the readers to comprehend the various radiation-induced effects and recognize how critically important radiation hardness is. The capacity to accurately detect and measure the radiation levels is a prerequisite for moving on to the next stage of the process, which is strategizing mitigation techniques. Consequently, a conversation about ionizing radiation cannot be considered in its entirety without mentioning the various radiation measuring or sensing elements. Therefore, this chapter also presents an overview of the numerous approaches and devices used in measuring radiation, primarily emphasizing semiconductor-based radiation sensors. In the end, the chapter summarizes several strategies and recommendations for reducing radiation-induced effects, with the implementation of radiation-hardened-by-design at the circuit level as the primary focus of the discussion.

Chapter 3 discusses the fundamentals of time-based signal processing. It provides an overview of architectures and briefly explains the working principles of various time-based building blocks frequently referred to as VTCs, TDCs, and DTCs. The chapter also discusses the advantages and disadvantages of time-based signal processing compared to traditional voltage-domain circuits. It touches on various challenges in terms of implementation. Time-based signal processing circuits are primarily built using pseudo-digital logic gates, and the waveforms look similar to digital signals with two logic levels. However, unlike digital circuits, time-based circuits operate with discrete analog values encoded using pulse widths or time differences, and those can also be represented as bipolar signals. Unlike voltage mode, time-based circuits take advantage of the decreasing feature sizes with increased time resolution and do away with static power usage with energy-efficient OTA-less designs. As discussed in this chapter, one of the basic building blocks is the VTC which converts the voltage-based signal into a time-based format based on the charging or discharging of a capacitor. The TDC is another well-known circuit. It uses delay cells as points of reference to digitize time-based data. DTCs generate pulse widths or time delays that can be changed digitally. Time registers and amplifiers are two of the other kinds of circuits that are utilized in time-based signal processing. Most time-based filter implementations, notably IIR and FIR filters, use these time-based arithmetic units. The accuracy of delay cells and time references is one of the primary concerns with time-based implementations. PLLs and DLLs are often used in time-based circuits to make low-jitter time references. However, these components make the system more complicated and use more energy. Time-based circuits appear to be significant contenders for voltage-based circuits thanks to better quantization capability in scaled CMOS processes. The statement, however, might seem inaccurate and limited to a few particular time-based circuit topologies. Smaller feature sizes achieved with scaling improve the timing skew and the gate delays, which are the primary factors limiting the quantization levels for Nyquist-rate TDCs with single-shot precision, even though the device noise contribution increases. However, for oversampling TDCs, achieving a small effective time resolution with reduced feature sizes might be challenging. The achievable dynamic range for such TDCs is constrained by the in-band noise rather than the quantization noise. Smaller device sizes increase the amount of device noise that goes into timing jitter, which raises the in-band noise levels. The situation may worsen with constrained power budgets for applications like sensor interfaces. There, the problem is exacerbated as the lower slew rate of transitions increases the timing jitter. A trade-off should be explored to optimize the in-band noise within the limited power-budget while maximizing energy efficiency.

The details of implementation and radiation assessment of two different LC-tank-based QVCOs (PQVCO and SQVCO) generating frequencies within the range of 2.5–2.9 GHz have been discussed in Chap. 4. The oscillators could generate quadrature phases based on parallel and super-harmonic coupling. As reported in the literature, this is the first comprehensive experimental study of TID-induced performance variations in LC-tank-based QVCOs implemented in 65 nm bulk CMOS technology. The measurement results and the detailed overview of the test setup for the radiation experiment have been provided in that chapter. The QVCOs were targeted for use inside PLLs and half-rate CDRs in application areas withstanding high TID levels, such as HEP experiments and deep-space probes with extended mission lifetimes. The working prototypes of both oscillators were evaluated experimentally under X-ray irradiation (TID <=100 Mrad (SiO$_2$)). The data obtained during irradiation were statistically analyzed to perform a comparative study of the QVCOs and evaluate the effectiveness of the RHBD techniques employed in the designs. Considering the cumulative dose effects on various performance parameters, the oscillators' most vulnerable parameter appeared to be the quadrature phase. The accuracy of the quadrature phase depends on the relative mismatch between the oscillator cores' tank capacitors. In the case of high-frequency (> 10 GHz) applications, the value of such tank capacitors would be less, which could aggravate the radiation-induced variations in the output phase.

The design and development of a novel $\Delta\Sigma$ CDC intended for use in HEP experiments have been discussed in Chap. 5. It has provided the specifics of the architecture and explained the operating principle of the numerous essential building components. In addition, the chapter has presented the findings from the measurements. The proposed architecture of CDC utilizes time-based circuit components to perform the arithmetic operations (addition, subtraction, integration, etc.) of time-based information. The proposed design follows a dual quantization-based MASH 1–0 architecture to accomplish $\Delta\Sigma$ modulation. The prototypes were implemented in a commercial 65 nm CMOS process and validated experimentally in a radiationless lab environment and under heavy-ion exposure (Xe-ion, Energy = 2059 MeV). The CDC prototypes could measure capacitance in the range of 0–3.75 pF and have achieved ENOB of 12.9 bits with an energy efficiency of 0.18 pJ/conversion step. The CDC accomplished the fastest conversion time compared to its counterparts reported so far in the literature, enabling it to be utilized with sensor arrays with high throughput. The design incorporates a foreground DLL-based calibration to accommodate TID-induced performance deterioration and to lengthen the lifespan of operation when subjected to radiation exposure. To protect against SEEs, the essential functional components of the proposed design were constructed with the help of TMR-based mitigation methods. The radiation hardness of the implemented prototypes has been characterized and evaluated up to a LET energy level of 65 MeV·cm^2/mg in the air at ambient temperature. One of the key research outcomes of this IC development project is the proof-of-concept demonstration of first-order $\Delta\Sigma$ modulation with a time-mode IIR filter implemented using time-based arithmetic circuit components. The numerous insights and learnings made throughout this project have laid the groundwork for designing higher order time-mode discrete-time loop filters, which can be used in quantization noise shaping during $\Delta\Sigma$ modulation.

The design and development of the proposed $\Delta\Sigma$-TDC intended for use in ultra-low power ADPLL interfaces have been discussed in Chap. 6. The implemented ASIC prototype demonstrated a single-loop higher order ($\leq$3) $\Delta\Sigma$-modulation using time-mode arithmetic units. The TDC core occupied an area of 0.018 mm^2 with a maximum measured T_{range} of 98 (= ±49) ns. One of the key design aspects of the TDC was the configurable noise transfer function with changeable filter order. The proposed architecture used a time-based FIR filter in feedback with configurable filter coefficients to realize first, second, and third orders of $\Delta\Sigma$ modulation. The proof-of-concept demonstration of reconfigurable filter coefficients would lead to a systematic design strategy for various architectures of $\Delta\Sigma$-type TDCs. With this approach, it would be feasible to accomplish an optimal noise transfer function with arbitrary filter coefficients, as in the case of well-known voltage-domain delta-sigma ADCs. The TDC prototypes have been implemented in 65 nm CMOS technology and operated with near-threshold supply voltage (0.6 V). To the author's knowledge, these TDC prototypes are the first implemented and validated designs of $\Delta\Sigma$ TDCs powered with sub-1V supplies. The power consumption of the TDC core was within the range of 11–15.8 μW with the sampling frequency of 10 MHz clock frequency for multiple orders ($\leq$3) of $\Delta\Sigma$-modulation. The timing skew and propagation delays of circuit components severely constrain the operating range of time-based circuits. The proposed architecture has circumvented the limitation and maximized the sampling range by employing a time-interleaved strategy to process the time-based information. The different circuit components of the $\Delta\Sigma$-TDC were optimized for dynamic power consumption. As measured, the implemented design achieved the best state-of-the-art energy efficiency with a range of 15.6–89.2 fJ/conversion step compared to the current higher order ($\leq$3) $\Delta\Sigma$-TDCs so far reported in the literature. For a limited power budget, the dynamic range of time-to-digital conversion using oversampling techniques gets limited by the in-band contribution of timing jitter from different circuit components. It gets worse with devices operating with near-threshold supplies. One of the key findings from the various measurement results of the proposed TDC samples is the concern for increased noise levels at the outputs of time-based circuits operated with reduced supply voltages.

Chapter 7 has discussed the implementation of a radiation-hardened, fractional-N ADPLL-based frequency synthesizer to generate different clock and pulse-width-modulated signals. Reconfigurability was made more accessible by the integrated ADPLL design, which also featured outputs compatible with LVDS, LVCMOS, and PWM formats. These features allowed the device to meet a wide variety of customer requirements. The ASIC prototype will likely be the first commercially available radiation-hardened frequency synthesizer targeted for future critical space missions and nuclear-related markets. An in-depth discussion of the ADPLL architecture has been presented in this chapter. Throughout this discussion, various circuit-level implementations have been elaborated upon, primarily emphasizing mitigation strategies implemented at the circuit level. The primary takeaways and numerous findings from the radiation study conducted during the first project (Chap. 4) served as the foundation upon which the radiation resilience of the DCOs was developed as a strategy. A scheme based on two LC cores has been utilized to ensure that the DCOs can satisfy the requirements for a wide tuning range. A novel extended multi-modulus-divider architecture equipped with TMR for radiation hardness has been implemented and optimized for dynamic power consumption. In addition, AC-coupled CMOS output buffers have been implemented, each with a duty-cycle adjustment mechanism built into it. The chapter also discussed the various performance metrics and the findings of the radiation assessment performed on the samples. The prototypes were implemented using a 65 nm CMOS process and hardened with respect to both TID and SEEs (such as SELs, SEUs, and SETs).

8.2 Research Valorization

The development of the IC prototype of a radiation-hardened ADPLL has been discussed in Chap. 7 and was implemented and verified experimentally as a part of an industrial collaboration between the ADVISE research group of KU Leuven and MAGICS Technologies. The project was supported financially by the ESA's ARTES Core Competitiveness Programme. As a part of research valorization in this project, the key contributions can be noted as the design of various radiation-hardened integrated circuit components such as wide-tuning range dual-core digitally controlled LC oscillators, novel dynamic power optimized multi-modulus divider with triple modular redundancy, and AC-coupled CMOS output buffers with duty-cycle correction mechanism.

The industrial ASIC prototype featured a radiation-hardened fractional-N frequency synthesizer for onboard clock generation (1 MHz–2.5 GHz) and to facilitate serial communication. The samples were validated experimentally to achieve a radiation tolerance of more than 100 krad (Si) in terms of TID levels and SEL immunity for LET levels up to 62.5 MeV.cm^2/mg. The designed prototypes were targeted for applications in various harsh radiation environment areas, such as in future nuclear power plants and high-energy physics experiments, and are ideally suited for use in satellites or deep-space explorations with an extended mission lifetime.

8.3 Suggestions for Future Work

Radiation Study of Multi-phase Clock Generators

Integrated transceivers play a crucial role in wireline as well as optical communication interfaces. These transceivers eventually require multi-phase clock signals to drive half-rate CDR circuits and mainly assist with data-slicing in serial-to-parallel conversion and vice versa. Nowadays, multi-tap FIR equalization with <1 UI is increasingly used for bandwidth enhancement and requires quadrature or eight-phase clock signals for filter implementation. In modern days, those transceivers have found indispensable use in implementing high-speed communication in space and during HEP experiments. Recently, there has been a growing interest from commercial entities to deploy satellite constellations for use in LEO and featuring in-space optical communications. Exposure to ionizing radiation poses one of the significant challenges for such endeavors, and there exists a need to characterize and verify the performance degradation of the transceivers. As a part of this research, two different architectures of VCOs generating quadrature phases were studied under radiation. There exist several other architectures to generate multi-phase clocks. Unlike QVCOs, injection-locked ring oscillators generating low-jitter appear as a viable alternative. Although delays in ring oscillators are highly vulnerable to radiation, the phase accuracy is less sensitive. It would be interesting to conduct a comparative study and assessment of radiation sensitivities of injection-locked ring oscillators and QVCOs under several operating conditions.

Time-based Circuits for Dosimetry

Ionizing radiation affects electronics as well as living beings. These days many healthcare personnel are subjected to radiation exposure (X-ray) throughout their careers due to various diagnostic and therapeutic procedures. Such exposure concerns extend to patients as well. Traditionally, working professionals in such healthcare sectors require to wear dosimeters that could provide accumulated dose readings and are sensitive to micro-dose levels. Active integrated radiation dosimeters feature faster readout electronics with memory, facilitating continuous monitoring to ensure rapid and helpful counteractive measures. These dosimeters require very high resolution, but energy efficiency poses a major challenge as, in most cases, such dosimeters are powered by a battery. Oversampling data converters using traditional voltage-based circuits are suitable for such applications requiring high resolution; however, the main disadvantage is the energy efficiency. Time-based circuits featuring OTA-less designs are a viable alternative for achieving high resolution in combination with high energy efficiency. The $\Delta\Sigma$-CDC architecture, as illustrated in Chap. 5, can easily be adapted in the future to be used in dosimetry for measuring dose rate and accumulated dose information.

Usually, the time delay, which is considered the primary time-based signal, changes with respect to radiation dose. While exposed to radiation, the time delay produced by a circuit showcases a memory effect and varies with respect to accumulated dose. Therefore under radiation dose, the time delay acts like an intrinsic "integrator" of the dose rate. This feature can be exploited to simplify design and reduce circuit components in the case of noise-shaping data converters, which could eventually lead to a smaller area and lower power. A possible solution can be proposed for future work as illustrated in Fig. 8.1.

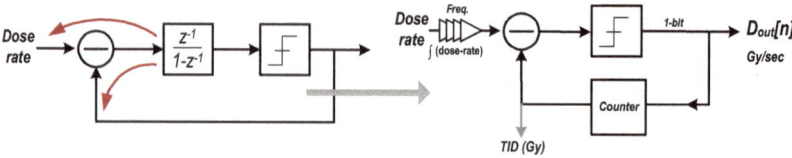

Fig. 8.1 Concept for time-based dosimeter

Ultra-Low-Power ADPLL Using Noise-Shaping TDCs

Frequency synthesizers based on ADPLL are widely used in modern-scaled CMOS technologies to implement SoCs and act as a backbone for wired and wireless communications. Ultra-low-power ADPLLs are also included in processor and memory interfaces for clock generation. Unlike analog PLLs, ADPLLs use TDCs for timing error measurement and DCOs for frequency generation. They are able to reduce the area and power dissipation in scaled technologies. Generally, ADPLL applications are targeted to achieve low jitter with a wide tuning range in combination with small area and power. These requirements present significant challenges in terms of circuit design, particularly in the presence of low-V_{DD}. The performance of an ADPLL greatly depends on the accuracy of the time-to-digital conversion, and the dissipated power in the TDCs affects the overall system's energy efficiency. Noise-shaping TDCs using time-based arithmetic units with high resolution appear as viable options for timing-error measurement in the closed-loop ADPLL circuits. The digital low-pass filter used in an ADPLL can be integrated with a noise-shaping TDC to act as a decimation filter. However, achieving high resolution with a small area and power is always difficult, and there exists a trade-off. As a part of this research, two time-based oversampling data converters based on time-based arithmetic units are implemented. The first one is the $\Delta\Sigma$-CDC which achieves very high resolution but at the expense of power. The second one is the $\Delta\Sigma$-CDC which is area-efficient and achieves a very good energy efficiency but at the expense of increased noise levels. These two designs lie at two extremes of the trade-off curve featuring resolution versus power trend. There are scopes to improve the design. A more suitable solution based on time-based arithmetic units can be explored in the future, which finds a balance between resolution and dissipated power while achieving good energy efficiency.

Bibliography

1. S. J. Gaul, N. van Vonno, S. H. Voldman, and W. H. Morris, *Integrated Circuit Design for Radiation Environments*. John Wiley & Sons, 2019.
2. J. L. Wirth and S. C. Rogers, "The Transient Response of Transistors and Diodes to Ionizing Radiation," *IEEE Transactions on Nuclear Science*, vol. 11, no. 5, pp. 24–38, 1964.
3. D. Binder, E. C. Smith, and A. B. Holman, "Satellite Anomalies from Galactic Cosmic Rays," *IEEE Transactions on Nuclear Science*, vol. 22, no. 6, pp. 2675–2680, 1975.
4. T. C. May and M. H. Woods, "A New Physical Mechanism for Soft Errors in Dynamic Memories," in *16th International Reliability Physics Symposium*, pp. 33–40, 1978.
5. "RADSAGA ITN Programme." https://radsaga.web.cern.ch/about-radsaga. Accessed: 12 October 2022.
6. A. Karmakar, J. Wang, J. Prinzie, V. De Smedt, and P. Leroux, "A Review of Semiconductor Based Ionising Radiation Sensors Used in Harsh Radiation Environments and Their Applications," *Radiation*, vol. 1, no. 3, pp. 194–217, 2021.
7. A. D. Rossin, "Dosimetry for Radiation Damage Studies," *IEEE Transactions on Nuclear Science*, vol. 11, no. 5, pp. 130–136, 1964.
8. R. M. Kloepper, "Neutron and Gamma-Ray Rate Sensitivities of Several Dynamic Detectors Used in Radiation Effects," *IEEE Transactions on Nuclear Science*, vol. 11, no. 5, pp. 137–144, 1964.
9. J. A. Zoutendyk and C. J. Malone, "Field Funneling and Range Straggling in Partially Depleted Silicon Surface-Barrier Detectors," *IEEE Transactions on Nuclear Science*, vol. 31, no. 6, pp. 1101–1105, 1984.
10. J. A. Zoutendyk, C. J. Malone, and L. S. Smith, " Experimental Determination of Single-Event Upset (SEU) as a Function of Collected Charge in Bipolar Integrated Circuits ," *IEEE Transactions on Nuclear Science*, vol. 31, no. 6, pp. 1167–1174, 1984.
11. J. Prinzie, F. M. Simanjuntak, P. Leroux, and T. Prodromakis, "Low-power Electronic Technologies for Harsh Radiation Environments," *Nature Electronics*, pp. 1–11, 2021.
12. F. Bellina, *Ionizing Radiation Effects in Nanoscale CMOS Technologies Exposed to Ultra-High Doses*. PhD thesis, University of Udine, 2018.

© The Editor(s) (if applicable) and The Author(s), under exclusive license to Springer Nature Switzerland AG 2024
A. Karmakar et al., *Integrated Time-Based Signal Processing Circuits for Harsh Radiation Environments*, Synthesis Lectures on Engineering, Science, and Technology, https://doi.org/10.1007/978-3-031-40620-1

13. H. Hughes, P. McMarr, M. Alles, E. Zhang, C. Arutt, B. Doris, D. Liu, R. Southwick, and P. Oldiges, "Total Ionizing Dose Radiation Effects on 14 nm FinFET and SOI UTBB Technologies," in *2015 IEEE Radiation Effects Data Workshop (REDW)*, pp. 1–6, IEEE, 2015.

14. D. Kobayashi, "Scaling Trends of Digital Single-Event Effects: A Survey of SEU and SET Parameters and Comparison With Transistor Performance," *IEEE Transactions on Nuclear Science*, vol. 68, no. 2, pp. 124–148, 2021.

15. T. Wang, K. Wang, L. Chen, A. Dinh, B. Bhuva, and R. Shuler, "A RHBD LC-Tank Oscillator Design Tolerant to Single-Event Transients," *IEEE Transactions on Nuclear Science*, vol. 57, no. 6, pp. 3620–3625, 2010.

16. S. Jagtap, D. Sharma, and S. Gupta, "Design of SET tolerant LC Oscillators Using Distributed Bias Circuitry," *Microelectronics Reliability*, vol. 55, no. 9-10, pp. 1537–1541, 2015.

17. J. Prinzie, J. Christiansen, P. Moreira, M. Steyaert, and P. Leroux, "Comparison of a 65 nm CMOS Ring- and LC-Oscillator Based PLL in Terms of TID and SEU Sensitivity," *IEEE Transactions on Nuclear Science*, vol. 64, no. 1, pp. 245–252, 2017.

18. S. Jagtap, S. Anmadwar, S. Rudrapati, and S. Gupta, "A Single-Event Transient-Tolerant High-Frequency CMOS Quadrature Phase Oscillator," *IEEE Transactions on Nuclear Science*, vol. 66, no. 9, pp. 2072–2079, 2019.

19. T. D. Loveless, S. Jagannathan, E. X. Zhang, D. M. Fleetwood, J. S. Kauppila, T. D. Haeffner, and L. W. Massengill, "Combined Effects of Total Ionizing Dose and Temperature on a K-Band Quadrature LC-Tank VCO in a 32 nm CMOS SOI Technology," *IEEE Transactions on Nuclear Science*, vol. 64, no. 1, pp. 204–211, 2017.

20. D. M. Fleetwood, "Radiation Effects in a Post-Moore World," *IEEE Transactions on Nuclear Science*, vol. 68, no. 5, pp. 509–545, 2021.

21. L. Palkuti, M. Alles, and H. Hughes, "The Role of Radiation Effects in SOI Technology Development," in *2014 SOI-3D-Subthreshold Microelectronics Technology Unified Conference (S3S)*, pp. 1–2, IEEE, 2014.

22. J. Riffaud, M. Gaillardin, C. Marcandella, N. Richard, O. Duhamel, M. Martinez, M. Raine, P. Paillet, T. Lagutere, F. Andrieu, *et al.*, "TID Response of Nanowire Field-Effect Transistors: Impact of the Back-Gate Bias," *IEEE Transactions on Nuclear Science*, vol. 67, no. 10, pp. 2172–2178, 2020.

23. J. M. Benedetto, H. E. Boesch, F. B. McLean, and J. P. Mize, "Hole Removal in Thin-Gate MOSFETs by Tunneling," *IEEE Transactions on Nuclear Science*, vol. 32, no. 6, pp. 3916–3920, 1985.

24. N. S. Saks, M. G. Ancona, and J. A. Modolo, "Generation of Interface States by Ionizing Radiation in Very Thin MOS Oxides," *IEEE Transactions on Nuclear Science*, vol. 33, no. 6, pp. 1185–1190, 1986.

25. L. T. Clark, W. E. Brown, C. S. Young-Sciortino, J. D. Butler, S. M. Guertin, K. E. Holbert, P. Bikkina, S. Bhanushali, M. Turowski, and A. Levy, "Total Ionizing Dose Impact on 22-nm FD-SOI Ring Oscillator Current and Frequency," *IEEE Transactions on Nuclear Science*, vol. 69, no. 12, pp. 2305–2313, 2022.

26. R. Zhang, Q. Zheng, J. Cui, Y. Li, X. Yu, W. Lu, and Q. Guo, "Bias Dependence of Total Ionizing Dose Response in UTBB FD-SOI Transistors," *IEEE Transactions on Nuclear Science*, vol. 69, no. 12, pp. 2314–2323, 2022.

27. M. M. Mahmoud, J. Prinzie, D. Söderström, K. Niskanen, V. Pouget, A. Cathelin, S. Clerc, and P. Leroux, "Impact of Aging Degradation on Heavy-Ion SEU Response of 28-nm UTBB FD-SOI Technology," *IEEE Transactions on Nuclear Science*, vol. 69, no. 8, pp. 1865–1875, 2022.

28. M. Menouni, M. Barbero, F. Bompard, S. Bonacini, D. Fougeron, R. Gaglione, A. Rozanov, P. Valerio, and A. Wang, "1-Grad Total Dose Evaluation of 65 nm CMOS Technology for the HL-LHC Upgrades," *Journal of Instrumentation*, vol. 10, no. 05, p. C05009, 2015.

29. N. Demaria, "The Impact of Microelectronics on High Energy Physics Innovation: The Role of 65 nm CMOS Technology on New Generation Particle Detectors," *Frontiers in Physics*, vol. 9, p. 629028, 2021.

30. G. Borghello, E. Lerario, F. Faccio, H. Koch, G. Termo, S. Michelis, F. Marquez, F. Palomo, and F. Muñoz, "Ionizing Radiation Damage in 65 nm CMOS Technology: Influence of Geometry, Bias and Temperature at Ultra-High Doses," *Microelectronics Reliability*, vol. 116, p. 114016, 2021.

31. P. E. Dodd, M. R. Shaneyfelt, J. R. Schwank, and J. A. Felix, "Current and Future Challenges in Radiation Effects on CMOS Electronics," *IEEE Transactions on Nuclear Science*, vol. 57, no. 4, pp. 1747–1763, 2010.

32. P. Dodd, M. Shaneyfelt, J. Felix, and J. Schwank, "Production and Propagation of Single-event Transients in High-speed Digital Logic ICs," *IEEE Transactions on Nuclear Science*, vol. 51, no. 6, pp. 3278–3284, 2004.

33. A. Baschirotto, P. Harpe, and K. A. Makinwa, *Next-Generation ADCs, High-Performance Power Management, and Technology Considerations for Advanced Integrated Circuits: Advances in Analog Circuit Design 2019*. Springer, 2020.

34. S. Ziabakhsh, G. Gagnon, and G. W. Roberts, "The Peak-SNR Performances of Voltage-mode Versus Time-mode Circuits," *IEEE Transactions on Circuits and Systems II: Express Briefs*, vol. 65, no. 12, pp. 1869–1873, 2018.

35. E. Stassinopoulos and J. Raymond, "The Space Radiation Environment for Electronics," *Proceedings of the IEEE*, vol. 76, no. 11, pp. 1423–1442, 1988.

36. J. Barth, C. Dyer, and E. Stassinopoulos, "Space, Atmospheric, and Terrestrial Radiation Environments," *IEEE Transactions on Nuclear Science*, vol. 50, no. 3, pp. 466–482, 2003.

37. E. Normand, "Single-event Effects in Avionics," *IEEE Transactions on Nuclear Science*, vol. 43, no. 2, pp. 461–474, 1996.

38. T. S. Dunn and J. L. Williams, "Comparison of Cobalt-60 and Electron Accelerators for Radiation Sterilization," *IEEE Transactions on Nuclear Science*, vol. 26, no. 1, pp. 1776–1783, 1979.

39. P. A. Piroue, "Radiation Detectors in High Energy Physics," *IEEE Transactions on Nuclear Science*, vol. 13, no. 5, pp. 54–59, 1966.

40. M. Bruzzi, "Radiation Damage in Silicon Detectors for High-Energy Physics Experiments," *IEEE Transactions on Nuclear Science*, vol. 48, no. 4, pp. 960–971, 2001.

41. T. Oldham and F. McLean, "Total Ionizing Dose Effects in MOS Oxides and Devices," *IEEE Transactions on Nuclear Science*, vol. 50, no. 3, pp. 483–499, 2003.

42. D. Fleetwood, T. Meisenheimer, and J. Scofield, "1/f Noise and Radiation Effects in MOS Devices," *IEEE Transactions on Electron Devices*, vol. 41, no. 11, pp. 1953–1964, 1994.

43. R. Baumann, "Radiation-induced Soft Errors in Advanced Semiconductor Technologies," *IEEE Transactions on Device and Materials Reliability*, vol. 5, no. 3, pp. 305–316, 2005.

44. J. Srour, C. Marshall, and P. Marshall, "Review of Displacement Damage Effects in Silicon Devices," *IEEE Transactions on Nuclear Science*, vol. 50, no. 3, pp. 653–670, 2003.

45. M. Moll, "Displacement Damage in Silicon Detectors for High Energy Physics," *IEEE Transactions on Nuclear Science*, vol. 65, no. 8, pp. 1561–1582, 2018.

46. J. Srour, C. Marshall, and P. Marshall, "Review of displacement damage effects in silicon devices," *IEEE Transactions on Nuclear Science*, vol. 50, no. 3, pp. 653–670, 2003.

47. W. Beezhold, D. Beutler, J. Garth, and P. Griffin, "A Review of the 40-Year History of the NSREC'S Dosimetry and Facilities session (1963-2003)," *IEEE Transactions on Nuclear Science*, vol. 50, no. 3, pp. 635–652, 2003.

48. D. L. Hester, D. D. Glower, and L. J. Overton, "Use of Ferroelectrics for Gamma-Ray Dosimetry," *IEEE Transactions on Nuclear Science*, vol. 11, no. 5, pp. 145–154, 1964.

49. R. L. Birdsall, D. Binder, and W. M. Peffley, "A Strain Gage Dosimeter for Pulsed Radiation Environment," *IEEE Transactions on Nuclear Science*, vol. 15, no. 6, pp. 346–349, 1968.

50. D. R. Sukis, "Thermoluminescent Properties of CaF2: Dy TLD'S," *IEEE Transactions on Nuclear Science*, vol. 18, no. 6, pp. 185–189, 1971.

51. T. F. Wrobel and R. A. Berger, "Silicon Calorimeter System for Gamma and Electron-Beam Radiation Dosimetry," *IEEE Transactions on Nuclear Science*, vol. 22, no. 6, pp. 2314–2318, 1975.

52. L. Adams and A. Holmes-Siedle, "The Development of an MOS Dosimetry Unit for Use in Space," *IEEE Transactions on Nuclear Science*, vol. 25, no. 6, pp. 1607–1612, 1978.

53. G. J. Brucker, E. G. Stassinopoulos, O. Van Gunten, L. S. August, and T. M. Jordan, "The Damage Equivalence of Electrons, Protons, and Gamma Rays in MOS Devices," *IEEE Transactions on Nuclear Science*, vol. 29, no. 6, pp. 1966–1969, 1982.

54. W. J. Stapor, L. S. August, D. H. Wilson, T. R. Oldham, and K. M. Murray, "Proton and Heavy-Ion Radiation Damage Studies in MOS Transistors," *IEEE Transactions on Nuclear Science*, vol. 32, no. 6, pp. 4399–4404, 1985.

55. A. Rosenfeld, "MOSFET Dosimetry on Modern Radiation Oncology Modalities," *Radiation protection dosimetry*, vol. 101, pp. 393–8, 02 2002.

56. F. Ravotti, "Dosimetry Techniques and Radiation Test Facilities for Total Ionizing Dose Testing," *IEEE Transactions on Nuclear Science*, vol. 65, no. 8, pp. 1440–1464, 2018.

57. S. Kasahara, T. Mitani, K. Ogasawara, T. Takashima, M. Hirahara, and K. Asamura, "Application of Single-sided Silicon Strip Detector to Energy and Charge State Measurements of Medium Energy Ions in Space," *Nuclear Instruments and Methods in Physics Research Section A: Accelerators, Spectrometers, Detectors and Associated Equipment*, vol. 603, no. 3, pp. 355–360, 2009.

58. M. Havranek, T. Benka, M. Hejtmanek, Z. Janoska, V. Kafka, J. Kopecek, M. Kuklova, M. Marcisovska, M. Marcisovsky, G. Neue, P. Svihra, L. Tomasek, P. Vancura, and V. Vrba, "MAPS sensor for Radiation Imaging Designed in 180 nm SOI CMOS Technology," *Journal of Instrumentation*, vol. 13, pp. C06004–C06004, jun 2018.

59. E. Damulira, M. N. S. Yusoff, A. F. Omar, and N. H. Mohd Taib, "A Review: Photonic Devices Used for Dosimetry in Medical Radiation," *Sensors*, vol. 19, no. 10, p. 2226, 2019.

60. L. H. Zhang, S. P. Platt, R. Edwards, and C. Allabush, "In-situ Neutron Dosimetry for Single-Event Effect Accelerated Testing," *IEEE Transactions on Nuclear Science*, vol. 56, no. 4, pp. 2070–2076, 2009.

61. Y. Duan, Y. Yao, Z. Li, J. Zhou, P. Huang, and W. Gao, "SENSROC12: A Four-Channel Binary-Output Front-End Readout ASIC for Si-PIN-Based Personal Dosimeters," *IEEE Transactions on Nuclear Science*, vol. 66, no. 8, pp. 1976–1983, 2019.

62. H. N. Becker, W. H. Farr, and D. Q. Zhu, "Radiation Response of Emerging High Gain, Low Noise Detectors," *IEEE Transactions on Nuclear Science*, vol. 54, no. 4, pp. 1129–1135, 2007.

63. G. Bertuccio, D. Puglisi, D. Macera, R. Di Liberto, M. Lamborizio, and L. Mantovani, "Silicon Carbide Detectors for in Vivo Dosimetry," *IEEE transactions on nuclear science*, vol. 61, no. 2, pp. 961–966, 2014.

64. A. Holmes-Siedle and L. Adams, "RADFET: A Review of the Use of Metal-Oxide-Silicon Devices as Integrating Dosimeters," *International Journal of Radiation Applications and Instrumentation. Part C. Radiation Physics and Chemistry*, vol. 28, no. 2, pp. 235–244, 1986.

65. D. Peet and M. Pryor, "Evaluation of a MOSFET Radiation Sensor for the Measurement of Entrance Surface Dose in Diagnostic Radiology," *The British journal of radiology*, vol. 72, no. 858, pp. 562–568, 1999.

66. J. Kassabov, N. Nedev, and N. Smirnov, "Radiation Dosimeter Based on Floating Gate MOS Transistor," *Radiation Effects and Defects in Solids*, vol. 116, no. 1-2, pp. 155–158, 1991.

67. N. Tarr, K. Shortt, Y. Wang, and I. Thomson, "A Sensitive, Temperature-compensated, Zero-bias Floating Gate MOSFET Dosimeter," *IEEE Transactions on Nuclear Science*, vol. 51, no. 3, pp. 1277–1282, 2004.

68. E. Garcia-Moreno, E. Isern, M. Roca, R. Picos, J. Font, J. Cesari, and A. Pineda, "Floating Gate CMOS Dosimeter With Frequency Output," *IEEE Transactions on Nuclear Science*, vol. 59, no. 2, pp. 373–378, 2012.

69. B. Chatterjee, C. Mousoulis, D.-H. Seo, A. Kumar, S. Maity, S. M. Scott, D. J. Valentino, D. T. Morisette, D. Peroulis, and S. Sen, "A Wearable Real-Time CMOS Dosimeter With Integrated Zero-Bias Floating Gate Sensor and an 861-nW 18-Bit Energy-Resolution Scalable Time-Based Radiation to Digital Converter," *IEEE Journal of Solid-State Circuits*, vol. 55, no. 3, pp. 650–665, 2020.

70. E. Garcia-Moreno, E. Isern, M. Roca, R. Picos, J. Font, J. Cesari, and A. Pineda, "Temperature Compensated Floating Gate MOS Radiation Sensor With Current Output," *IEEE Transactions on Nuclear Science*, vol. 60, no. 5, pp. 4026–4030, 2013.

71. E. Pikhay, Y. Roizin, and Y. Nemirovsky, "Ultra-low Power Consuming Direct Radiation Sensors Based on Floating Gate Structures," *Journal of Low Power Electronics and Applications*, vol. 7, no. 3, p. 20, 2017.

72. J. Prinzie, S. Thys, B. Van Bockel, J. Wang, V. De Smedt, and P. Leroux, "An SRAM-Based Radiation Monitor With Dynamic Voltage Control in 0.18- μ m CMOS Technology," *IEEE Transactions on Nuclear Science*, vol. 66, no. 1, pp. 282–289, 2019.

73. R. Harboe-Sorensen, F.-X. Guerre, and A. Roseng, "Design, Testing and Calibration of a Reference SEU Monitor System," in *2005 8th European Conference on Radiation and Its Effects on Components and Systems*, pp. B3–1–B3–7, 2005.

74. G. Spiezia, P. Peronnard, A. Masi, M. Brugger, M. Brucoli, S. Danzeca, R. G. Alia, R. Losito, J. Mekki, P. Oser, R. Gaillard, and L. Dusseau, "A New RadMon Version for the LHC and its Injection Lines," *IEEE Transactions on Nuclear Science*, vol. 61, no. 6, pp. 3424–3431, 2014.

75. R. Harboe-Sorensen, C. Poivey, N. Fleurinck, K. Puimege, A. Zadeh, F.-X. Guerre, F. Lochon, M. Kaddour, L. Li, D. Walter, A. Keating, A. Jaksic, and M. Poizat, "The Technology Demonstration Module On-Board PROBA-II," *IEEE Transactions on Nuclear Science*, vol. 58, no. 3, pp. 1001–1007, 2011.

76. R. Secondo, G. Foucard, S. Danzeca, R. Losito, P. Peronnard, A. Masi, M. Brugger, and L. Dusseau, "Embedded Detection and Correction of SEU Bursts in SRAM Memories Used as Radiation Detectors," *IEEE Transactions on Nuclear Science*, vol. 63, no. 4, pp. 2168–2175, 2016.

77. M. Brucoli, S. Danzeca, M. Brugger, A. Masi, A. Pineda, J. Cesari, L. Dusseau, and F. Wrobel, "Floating Gate Dosimeter Suitability for Accelerator-Like Environments," *IEEE Transactions on Nuclear Science*, vol. 64, no. 8, pp. 2054–2060, 2017.

78. S. Gerardin, M. Bagatin, A. Paccagnella, V. Ferlet-Cavrois, A. Visconti, and C. D. Frost, "Neutron and Alpha Single Event Upsets in Advanced NAND Flash Memories," *IEEE Transactions on Nuclear Science*, vol. 61, no. 4, pp. 1799–1805, 2014.

79. M. Bagatin, S. Gerardin, A. Paccagnella, S. Beltrami, E. Camerlenghi, M. Bertuccio, A. Costantino, A. Zadeh, V. Ferlet-Cavrois, G. Santin, and E. Daly, "Effects of Heavy-Ion Irradiation on Vertical 3-D NAND Flash Memories," *IEEE Transactions on Nuclear Science*, vol. 65, no. 1, pp. 318–325, 2018.

80. D. Chen, E. Wilcox, R. L. Ladbury, C. Seidleck, H. Kim, A. Phan, and K. A. LaBel, "Heavy Ion and Proton-Induced Single Event Upset Characteristics of a 3-D NAND Flash Memory," *IEEE Transactions on Nuclear Science*, vol. 65, no. 1, pp. 19–26, 2018.

81. M. Bagatin, S. Gerardin, A. Paccagnella, S. Beltrami, C. Cazzaniga, and C. D. Frost, "Atmospheric Neutron Soft Errors in 3-D NAND Flash Memories," *IEEE Transactions on Nuclear Science*, vol. 66, no. 7, pp. 1361–1367, 2019.

82. M. Andjelkovic, J. Chen, A. Simevski, Z. Stamenkovic, M. Krstic, and R. Kraemer, "A Review of Particle Detectors for Space-Borne Self-Adaptive Fault-Tolerant Systems," in *2020 IEEE East-West Design Test Symposium (EWDTS)*, pp. 1–8, 2020.

83. Y. Li, W. M. Porter, R. Ma, M. A. Reynolds, B. J. Gerbi, and S. J. Koester, "Capacitance-Based Dosimetry of Co-60 Radiation Using Fully-Depleted Silicon-on-Insulator Devices," *IEEE Transactions on Nuclear Science*, vol. 62, no. 6, pp. 3012–3019, 2015.

84. Y. Li, V. R. S. K. Chaganti, M. A. Reynolds, B. J. Gerbi, and S. J. Koester, " Demonstration of a Passive Wireless Radiation Detector Using Fully-Depleted Silicon-on-Insulator Variable Capacitors ," *IEEE Transactions on Nuclear Science*, vol. 64, no. 1, pp. 544–549, 2017.

85. G. Vaidhyanathan and S. J. Koester, "High-Q FDSOI Varactors for Wireless Radiation Sensing," in *IEEE 2011 International SOI Conference*, pp. 1–2, 2011.

86. R. Baumann and K. Kruckmeyer, "Radiation Handbook for Electronics," tech. rep., Texas Instruments: Dallas, TX, USA, 2019.

87. F. Faccio, S. Michelis, D. Cornale, A. Paccagnella, and S. Gerardin, "Radiation-Induced Short Channel (RISCE) and Narrow Channel (RINCE) Effects in 65 and 130 nm MOSFETs," *IEEE Transactions on Nuclear Science*, vol. 62, no. 6, pp. 2933–2940, 2015.

88. F. Yuan, *CMOS Time-mode Circuits and Systems: Fundamentals and Applications*. CRC Press, 2018.

89. K. Kim, W. Yu, and S. Cho, "A 9 bit, 1.12 ps Resolution 2.5 b/Stage Pipelined Time-to-Digital Converter in 65 nm CMOS Using Time-Register," *IEEE Journal of Solid-State Circuits*, vol. 49, no. 4, pp. 1007–1016, 2014.

90. K. Zhu, J. Feng, and Y. Lyu, "A 336fs rms 0.89 mW 200MS/s 5MHz Bandwidth 2–2 MASH Δ Σ Time-to-Digital Converter with Differential Time-Mode Arithmetic Units," in *2020 IEEE International Symposium on Circuits and Systems (ISCAS)*, pp. 1–4, IEEE, 2020.

91. Y. Wu, P. Lu, and R. B. Staszewski, "A Time-Domain 147fsrms 2.5-MHz Bandwidth Two-Step Flash-Mash 1-1-1 Time-to-Digital Converter With Third-Order Noise-Shaping and Mismatch Correction," *IEEE Transactions on Circuits and Systems I: Regular Papers*, vol. 67, no. 8, pp. 2532–2545, 2020.

92. S. Zhu, B. Xu, B. Wu, K. Soppimath, and Y. Chiu, "A Skew-Free 10 GS/s 6 bit CMOS ADC With Compact Time-Domain Signal Folding and Inherent DEM," *IEEE Journal of Solid-State Circuits*, vol. 51, no. 8, pp. 1785–1796, 2016.

93. A. R. Macpherson, J. W. Haslett, and L. Belostotski, "A 5GS/s 4-bit Time-based Single-channel CMOS ADC for Radio Astronomy," in *Proceedings of the IEEE 2013 Custom Integrated Circuits Conference*, pp. 1–4, 2013.

94. M. Park and M. H. Perrott, "A Single-slope 80MS/s ADC using Two-Step Time-to-Digital Conversion," in *2009 IEEE International Symposium on Circuits and Systems*, pp. 1125–1128, 2009.

95. L.-J. Chen and S.-I. Liu, "A 12-bit 3.4 MS/s Two-Step Cyclic Time-Domain ADC in 0.18-μm CMOS," *IEEE Transactions on Very Large Scale Integration (VLSI) Systems*, vol. 24, no. 4, pp. 1470–1483, 2016.

96. V. Dhanasekaran, M. Gambhir, M. M. Elsayed, E. Sanchez-Sinencio, J. Silva-Martinez, C. Mishra, L. Chen, and E. Pankratz, "A 20MHz BW 68dB DR CT ΔΣ ADC based on a

multi-bit time-domain quantizer and feedback element," in *2009 IEEE International Solid-State Circuits Conference - Digest of Technical Papers*, pp. 174–175,175a, 2009.

97. M. M. Elsayed, V. Dhanasekaran, M. Gambhir, J. Silva-Martinez, and E. Sanchez-Sinencio, "A 0.8 ps DNL Time-to-Digital Converter With 250 MHz Event Rate in 65 nm CMOS for Time-Mode-Based $\Sigma\Delta$ Modulator," *IEEE Journal of Solid-State Circuits*, vol. 46, no. 9, pp. 2084–2098, 2011.

98. W. Jung, Y. Mortazavi, B. L. Evans, and A. Hassibi, "An all-digital PWM-based $\Delta\Sigma$ ADC with an inherently matched multi-bit quantizer," in *Proceedings of the IEEE 2014 Custom Integrated Circuits Conference*, pp. 1–4, 2014.

99. R. Enomoto, T. Iizuka, T. Koga, T. Nakura, and K. Asada, "A 16-bit 2.0-ps Resolution Two-Step TDC in 0.18-μm CMOS Utilizing Pulse-Shrinking Fine Stage With Built-In Coarse Gain Calibration," *IEEE Transactions on Very Large Scale Integration (VLSI) Systems*, vol. 27, no. 1, pp. 11–19, 2019.

100. P. Chen, S.-L. Liu, and J. Wu, "A CMOS Pulse-Shrinking Delay Element for Time Interval Measurement," *IEEE Transactions on Circuits and Systems II: Analog and Digital Signal Processing*, vol. 47, no. 9, pp. 954–958, 2000.

101. A. Mantyniemi, T. Rahkonen, and J. Kostamovaara, "A CMOS Time-to-Digital Converter (TDC) Based On a Cyclic Time Domain Successive Approximation Interpolation Method," *IEEE Journal of Solid-State Circuits*, vol. 44, no. 11, pp. 3067–3078, 2009.

102. D. J. Lee, F. Yuan, and Y. Zhou, "All-Digital Successive Approximation TDC in Time-Mode Signal Processing," in *2021 IEEE International Symposium on Circuits and Systems (ISCAS)*, pp. 1–4, 2021.

103. J.-S. Kim, Y.-H. Seo, Y. Suh, H.-J. Park, and J.-Y. Sim, "A 300-MS/s, 1.76-ps-Resolution, 10-b Asynchronous Pipelined Time-to-Digital Converter With on-Chip Digital Background Calibration in 0.13-μm CMOS," *IEEE Journal of Solid-State Circuits*, vol. 48, no. 2, pp. 516–526, 2013.

104. Y. Wu, M. Shahmohammadi, Y. Chen, P. Lu, and R. B. Staszewski, "A 3.5-6.8-GHz Wide-Bandwidth DTC-Assisted Fractional-N All-Digital PLL With a MASH $\Delta\Sigma$ -TDC for Low In-Band Phase Noise," *IEEE Journal of Solid-State Circuits*, vol. 52, no. 7, pp. 1885–1903, 2017.

105. B. Liu, Y. Zhang, J. Qiu, H. C. Ngo, W. Deng, K. Nakata, T. Yoshioka, J. Emmei, J. Pang, A. T. Narayanan, H. Zhang, T. Someya, A. Shirane, and K. Okada, "A Fully Synthesizable Fractional-N MDLL With Zero-Order Interpolation-Based DTC Nonlinearity Calibration and Two-Step Hybrid Phase Offset Calibration," *IEEE Transactions on Circuits and Systems I: Regular Papers*, vol. 68, no. 2, pp. 603–616, 2021.

106. A. Elmallah, M. G. Ahmed, A. Elkholy, W.-S. Choi, and P. K. Hanumolu, "A 1.6ps peak-INL 5.3ns Range Two-Step Digital-to-Time Converter in 65nm CMOS," in *2018 IEEE Custom Integrated Circuits Conference (CICC)*, pp. 1–4, 2018.

107. J. Z. Ru, C. Palattella, P. Geraedts, E. Klumperink, and B. Nauta, "A High-Linearity Digital-to-Time Converter Technique: Constant-Slope Charging," *IEEE Journal of Solid-State Circuits*, vol. 50, no. 6, pp. 1412–1423, 2015.

108. P. Chen, F. Zhang, Z. Zong, S. Hu, T. Siriburanon, and R. B. Staszewski, "A 31- μ W, 148-fs Step, 9-bit Capacitor-DAC-Based Constant-Slope Digital-to-Time Converter in 28-nm CMOS," *IEEE Journal of Solid-State Circuits*, vol. 54, no. 11, pp. 3075–3085, 2019.

109. A. Karmakar, V. De Smedt, and P. Leroux, "TID Sensitivity Assessment of Quadrature LC-Tank VCOs Implemented in 65-nm CMOS Technology," *Electronics*, vol. 11, no. 9, p. 1399, 2022.

110. A. Musa, R. Murakami, T. Sato, W. Chaivipas, K. Okada, and A. Matsuzawa, "A Low Phase Noise Quadrature Injection Locked Frequency Synthesizer for MM-Wave Applications," *IEEE Journal of Solid-State Circuits*, vol. 46, no. 11, pp. 2635–2649, 2011.

111. I. R. Chamas and S. Raman, "Analysis and Design of a CMOS Phase-Tunable Injection-Coupled LC Quadrature VCO (PTIC-QVCO)," *IEEE Journal of Solid-State Circuits*, vol. 44, no. 3, pp. 784–796, 2009.

112. A. Eroglu, "Non-Invasive Quadrature Modulator Balancing Method to Optimize Image Band Rejection," *IEEE Transactions on Circuits and Systems I: Regular Papers*, vol. 61, no. 2, pp. 600–612, 2014.

113. A. Nikoofard, S. Kananian, and A. Fotowat-Ahmady, "A Fully Analog Calibration Technique for Phase and Gain Mismatches in Image-Reject Receivers," *AEU-International Journal of Electronics and Communications*, vol. 69, no. 5, pp. 823–835, 2015.

114. J. Lee and B. Razavi, "A 40-Gb/s Clock and Data Recovery Circuit in 0.18-μm CMOS Tchnology," *IEEE Journal of Solid-State Circuits*, vol. 38, no. 12, pp. 2181–2190, 2003.

115. A. Mazzanti, F. Svelto, and P. Andreani, "On the Amplitude and Phase Errors of Quadrature LC-tank CMOS Oscillators," *IEEE Journal of Solid-State Circuits*, vol. 41, no. 6, pp. 1305–1313, 2006.

116. D. Monda, G. Ciarpi, and S. Saponara, "Design and Verification of a 6.25 GHz LC-Tank VCO Integrated in 65 nm CMOS Technology Operating up to 1 Grad TID," *IEEE Transactions on Nuclear Science*, vol. 68, no. 10, pp. 2524–2532, 2021.

117. A. Karthigeyan, S. Radha, and E. Manikandan, "Single Event Transient Mitigation Techniques for a Cross-coupled LC Oscillator Including a Single-Event Transient Hardened CMOS LC-VCO Circuit," *IET Circuits, Devices & Systems*, vol. 16, no. 2, pp. 178–188, 2022.

118. A. Nikoofard, S. Kananian, and A. Fotowat-Ahmady, "Off-Resonance Oscillation, Phase Retention, and Orthogonality Modeling in Quadrature Oscillators," *IEEE Transactions on Circuits and Systems I: Regular Papers*, vol. 63, no. 6, pp. 883–894, 2016.

119. L. Zhang, N.-C. Kuo, and A. M. Niknejad, "A 37.5-45 GHz Superharmonic-Coupled QVCO With Tunable Phase Accuracy in 28 nm CMOS," *IEEE Journal of Solid-State Circuits*, vol. 54, no. 10, pp. 2754–2764, 2019.

120. L. B. Oliveira, J. R. Fernandes, I. M. Filanovsky, C. J. Verhoeven, and M. M. Silva, *Analysis and Design of Quadrature Oscillators*. Springer Science & Business Media, 2008.

121. J. Koo, B. Kim, H.-J. Park, and J.-Y. Sim, "A Quadrature RC Oscillator With Noise Reduction by Voltage Swing Control," *IEEE Transactions on Circuits and Systems I: Regular Papers*, vol. 66, no. 8, pp. 3077–3088, 2019.

122. L. Dai and R. Harjani, "A Low-Phase-Noise CMOS Ring Oscillator With Differential Control and Quadrature Outputs," in *Proceedings 14th Annual IEEE International ASIC/SOC Conference (IEEE Cat. No.01TH8558)*, pp. 134–138, 2001.

123. E. J. Pankratz and E. Sanchez-Sinencio, "Multiloop High-Power-Supply-Rejection Quadrature Ring Oscillator," *IEEE Journal of Solid-State Circuits*, vol. 47, no. 9, pp. 2033–2048, 2012.

124. F. Behbahani, Y. Kishigami, J. Leete, and A. Abidi, "CMOS Mixers and Polyphase Filters for Large Image Rejection," *IEEE Journal of Solid-State Circuits*, vol. 36, no. 6, pp. 873–887, 2001.

125. S. P. Sah, X. Yu, and D. Heo, "Design and Analysis of a Wideband 15-35-GHz Quadrature Phase Shifter With Inductive Loading," *IEEE Transactions on Microwave Theory and Techniques*, vol. 61, no. 8, pp. 3024–3033, 2013.

126. T. H. Lee, *The Design of CMOS Radio-Frequency Integrated Circuits*. Cambridge university press, 2003.

127. P. Andreani, A. Bonfanti, L. Romano, and C. Samori, "Analysis and Design of a 1.8-GHz CMOS LC Quadrature VCO," *IEEE Journal of Solid-State Circuits*, vol. 37, no. 12, pp. 1737–1747, 2002.

128. H.-Y. Chang and Y.-T. Chiu, "K-Band CMOS Differential and Quadrature Voltage-Controlled Oscillators for Low Phase-Noise and Low-Power Applications," *IEEE Transactions on Microwave Theory and Techniques*, vol. 60, no. 1, pp. 46–59, 2012.

129. H.-Y. Chang, C.-H. Lin, Y.-C. Liu, Y.-L. Yeh, K. Chen, and S.-H. Wu, "65-nm CMOS Dual-Gate Device for Ka-Band Broadband Low-Noise Amplifier and High-Accuracy Quadrature Voltage-Controlled Oscillator," *IEEE Transactions on Microwave Theory and Techniques*, vol. 61, no. 6, pp. 2402–2413, 2013.

130. U. Decanis, A. Ghilioni, E. Monaco, A. Mazzanti, and F. Svelto, "A Low-Noise Quadrature VCO Based on Magnetically Coupled Resonators and a Wideband Frequency Divider at Millimeter Waves," *IEEE Journal of Solid-State Circuits*, vol. 46, no. 12, pp. 2943–2955, 2011.

131. N.-C. Kuo, J.-C. Chien, and A. M. Niknejad, "Design and Analysis on Bidirectionally and Passively Coupled QVCO With Nonlinear Coupler," *IEEE Transactions on Microwave Theory and Techniques*, vol. 63, no. 4, pp. 1130–1141, 2015.

132. I. R. Chamas and S. Raman, "A Comprehensive Analysis of Quadrature Signal Synthesis in Cross-Coupled RF VCOs," *IEEE Transactions on Circuits and Systems I: Regular Papers*, vol. 54, no. 4, pp. 689–704, 2007.

133. A. Mirzaei, M. E. Heidari, R. Bagheri, S. Chehrazi, and A. A. Abidi, "The Quadrature LC Oscillator: A Complete Portrait Based on Injection Locking," *IEEE Journal of Solid-State Circuits*, vol. 42, no. 9, pp. 1916–1932, 2007.

134. J. Prinzie, J. Christiansen, P. Moreira, M. Steyaert, and P. Leroux, "A 2.56-GHz SEU Radiation Hard LC-tank VCO for High-Speed Communication Links in 65-nm CMOS Technology," *IEEE Transactions on Nuclear Science*, vol. 65, no. 1, pp. 407–412, 2017.

135. H. Tong, S. Cheng, Y.-C. Lo, A. I. Karsilayan, and J. Silva-Martinez, "An LC Quadrature VCO Using Capacitive Source Degeneration Coupling to Eliminate Bi-Modal Oscillation," *IEEE Transactions on Circuits and Systems I: Regular Papers*, vol. 59, no. 9, pp. 1871–1879, 2012.

136. S. Gierkink, S. Levantino, R. Frye, C. Samori, and V. Boccuzzi, "A Low-Phase-Noise 5-GHz CMOS Quadrature VCO using Superharmonic Coupling," *IEEE Journal of Solid-State Circuits*, vol. 38, no. 7, pp. 1148–1154, 2003.

137. E. Hegazi, H. Sjoland, and A. Abidi, "A Filtering Technique to Lower LC Oscillator Phase Noise," *IEEE Journal of Solid-State Circuits*, vol. 36, no. 12, pp. 1921–1930, 2001.

138. T. Wu, U.-K. Moon, and K. Mayaram, "Dependence of LC VCO Oscillation Frequency on Bias Current," in *2006 IEEE International Symposium on Circuits and Systems (ISCAS)*, pp. 4–5, 2006.

139. F. Heiman and G. Warfield, "The Effects of Oxide Traps on the MOS Capacitance," *IEEE Transactions on Electron Devices*, vol. 12, no. 4, pp. 167–178, 1965.

140. P. Fernández-Martínez, I. Cortés, S. Hidalgo, D. Flores, and F. R. Palomo, "Simulation of Total Ionising Dose in MOS capacitors," in *Proceedings of the 8th Spanish Conference on Electron Devices, CDE'2011*, pp. 1–4, 2011.

141. J. Rael and A. A. Abidi, "Physical Processes of Phase Noise in Differential LC Oscillators," in *Proceedings of the IEEE 2000 Custom Integrated Circuits Conference (Cat. No. 00CH37044)*, pp. 569–572, IEEE, 2000.

142. V. Re, L. Gaioni, M. Manghisoni, L. Ratti, and G. Traversi, "Comprehensive Study of Total Ionizing Dose Damage Mechanisms and Their Effects on Noise Sources in a 90 nm CMOS Technology," *IEEE Transactions on Nuclear Science*, vol. 55, no. 6, pp. 3272–3279, 2008.

143. G. Mazza and S. Panati, "A Compact, Low Jitter, CMOS 65 nm 4.8–6 GHz Phase-Locked Loop for Applications in HEP Experiments Front-End Electronics," *IEEE Transactions on Nuclear Science*, vol. 65, no. 5, pp. 1212–1217, 2018.

144. S. Biereigel, S. Kulis, P. Leitao, R. Francisco, P. Moreira, P. Leroux, and J. Prinzie, "A Low Noise Fault Tolerant Radiation Hardened 2.56 Gbps Clock-Data Recovery Circuit With High

Speed Feed Forward Correction in 65 nm CMOS," *IEEE Transactions on Circuits and Systems I: Regular Papers*, vol. 67, no. 5, pp. 1438–1446, 2019.

145. A. Karmakar, V. De Smedt, and P. Leroux, "Pseudo-Differential Time-Domain Integrator Using Charge-Based Time-Domain Circuits," in *2021 IEEE 12th Latin America Symposium on Circuits and System (LASCAS)*, pp. 1–4, 2021.

146. A. Karmakar, V. De Smedt, and P. Leroux, "A 0.18 pJ/Step Time-Domain 1st Order $\Delta\Sigma$ Capacitance-to-Digital Converter in 65-nm CMOS," in *2021 IEEE International Symposium on Circuits and Systems (ISCAS)*, pp. 1–5, 2021.

147. S. Park, G.-H. Lee, and S. Cho, "A 2.92 μW Capacitance-to-Digital Converter With Differential Bondwire Accelerometer, On-Chip Air Pressure, and Humidity Sensor in 0.18-μm CMOS," *IEEE Journal of Solid-State Circuits*, vol. 54, no. 10, pp. 2845–2856, 2019.

148. U. Ferlito, A. D. Grasso, S. Pennisi, M. Vaiana, and G. Bruno, "Sub-Femto-Farad Resolution Electronic Interfaces for Integrated Capacitive Sensors: A Review," *IEEE Access*, vol. 8, pp. 153969–153980, 2020.

149. H. Xu, X. Liu, and L. Yin, "A Closed-Loop $\Sigma\Delta$ Interface for a High-Q Micromechanical Capacitive Accelerometer With 200 ng/$\sqrt{}$Hz Input Noise Density," *IEEE Journal of Solid-State Circuits*, vol. 50, no. 9, pp. 2101–2112, 2015.

150. Y. Wang, Q. Fu, Y. Zhang, W. Zhang, D. Chen, L. Yin, and X. Liu, "A Digital Closed-loop Sense MEMS Disk Resonator Gyroscope Circuit Design Based on Integrated Analog Front-end," *Sensors*, vol. 20, no. 3, p. 687, 2020.

151. X. Tang, S. Li, X. Yang, L. Shen, W. Zhao, R. P. Williams, J. Liu, Z. Tan, N. A. Hall, D. Z. Pan, *et al.*, "An Energy-efficient Time-domain Incremental Zoom Capacitance-to-Digital Converter," *IEEE Journal of Solid-State Circuits*, vol. 55, no. 11, pp. 3064–3075, 2020.

152. H.-J. Kwon, J.-s. Lee, J.-Y. Sim, and H.-J. Park, "A High-Gain Wide-Input-Range Time Amplifier With an Open-Loop Architecture and a Gain Equal to Current Bias Ratio," in *IEEE Asian Solid-State Circuits Conference 2011*, pp. 325–328, IEEE, 2011.

153. A. Sanyal and N. Sun, "An Energy-Efficient Hybrid SAR-VCO $\Delta\Sigma$ Capacitance-to-Digital Converter in 40-nm CMOS," *IEEE Journal of Solid-State Circuits*, vol. 52, no. 7, pp. 1966–1976, 2017.

154. A. Alhoshany and K. N. Salama, "A Precision, Energy-efficient, Oversampling, Noise-shaping Differential SAR Capacitance-to-Digital Converter," *IEEE Transactions on Instrumentation and Measurement*, vol. 68, no. 2, pp. 392–401, 2018.

155. B. Li, W. Wang, J. Liu, W.-J. Liu, Q. Yang, and W.-B. Ye, "A 1 pF-to-10 nF Generic Capacitance-to-Digital Converter Using Zero-Crossing $\Delta\Sigma$ Modulation," *IEEE Transactions on Circuits and Systems I: Regular Papers*, vol. 65, no. 7, pp. 2169–2182, 2017.

156. Y. He, Z.-y. Chang, L. Pakula, S. H. Shalmany, and M. Pertijs, "27.7 A 0.05 mm 2 1V Capacitance-to-Digital Converter Based on Period Modulation," in *2015 IEEE International Solid-State Circuits Conference-(ISSCC) Digest of Technical Papers*, pp. 1–3, IEEE, 2015.

157. A. Elshazly, S. Rao, B. Young, and P. K. Hanumolu, "A 13B 315fsrms 2mW 500MS/s 1MHz Bandwidth Highly Digital Time-to-Digital Converter Using Switched Ring Oscillators," in *2012 IEEE International Solid-State Circuits Conference*, pp. 464–466, 2012.

158. W. Yu, K. Kim, and S. Cho, "A 148fs_{rms} Integrated Noise 4 MHz Bandwidth Second-Order $\Delta\Sigma$ Time-to-Digital Converter With Gated Switched-Ring Oscillator," *IEEE Transactions on Circuits and Systems I: Regular Papers*, vol. 61, no. 8, pp. 2281–2289, 2014.

159. M. B. Dayanik and M. P. Flynn, "Digital Fractional- N PLLs Based on a Continuous-Time Third-Order Noise-Shaping Time-to-Digital Converter for a 240-GHz FMCW Radar System," *IEEE Journal of Solid-State Circuits*, vol. 53, no. 6, pp. 1719–1730, 2018.

160. M. Strackx, B. Van Bockel, A. Karmakar, S. Ali, B. Boons, R. Van Dyck, H. Marien, Y. Cao, P. Leroux, and J. Prinzie, "A Fully Integrated 1 MHz–2.5 GHz Radiation-Hardened All-Digital

Frequency Synthesizer," in *2021 21th European Conference on Radiation and Its Effects on Components and Systems (RADECS)*, pp. 1–4, IEEE, 2021.

161. M. Mestice, B. Neri, G. Ciarpi, and S. Saponara, "Analysis and Design of Integrated Blocks for a 6.25 GHz Spacefibre PLL," *Sensors*, vol. 20, no. 14, p. 4013, 2020.

162. T. Loveless, L. Massengill, B. Bhuva, W. Holman, R. Reed, D. McMorrow, J. Melinger, and P. Jenkins, "A Single-Event-Hardened Phase-locked Loop Fabricated in 130 nm CMOS," *IEEE transactions on nuclear science*, vol. 54, no. 6, pp. 2012–2020, 2007.

163. S. Biereigel, S. Kulis, P. Moreira, A. Kölpin, P. Leroux, and J. Prinzie, "Radiation-Tolerant All-Digital PLL/CDR with Varactorless LC DCO in 65 nm CMOS," *Electronics*, vol. 10, no. 22, p. 2741, 2021.

164. R. B. Staszewski, J. L. Wallberg, S. Rezeq, C.-M. Hung, O. E. Eliezer, S. K. Vemulapalli, C. Fernando, K. Maggio, R. Staszewski, N. Barton, *et al.*, "All-digital PLL and Transmitter for Mobile Phones," *IEEE journal of Solid-State circuits*, vol. 40, no. 12, pp. 2469–2482, 2005.

165. B. Van Bockel, *Radiation-Tolerant Sub-Picosecond Time-to-Digital Conversion.* PhD thesis, KU Leuven, 2023.

166. R. Staszewski, D. Leipold, K. Muhammad, and P. Balsara, "Digitally Controlled Oscillator (DCO)-based Architecture for RF Frequency Synthesis in a Deep-submicrometer CMOS Process," *IEEE Transactions on Circuits and Systems II: Analog and Digital Signal Processing*, vol. 50, no. 11, pp. 815–828, 2003.

167. C. Vaucher, I. Ferencic, M. Locher, S. Sedvallson, U. Voegeli, and Z. Wang, "A Family of Low-Power Truly Modular Programmable Dividers in Standard 0.35-μm CMOS Technology," *IEEE Journal of Solid-State Circuits*, vol. 35, no. 7, pp. 1039–1045, 2000.

168. A. Elkholy, S. Saxena, R. K. Nandwana, A. Elshazly, and P. K. Hanumolu, "A 2.0-5.5 GHz Wide Bandwidth Ring-Based Digital Fractional-N PLL With Extended Range Multi-Modulus Divider," *IEEE Journal of Solid-State Circuits*, vol. 51, no. 8, pp. 1771–1784, 2016.

169. K. Moustakas, P. Rymaszewski, T. Hemperek, H. Krüger, M. Vogt, T. Wang, and N. Wermes, " A Clock and Data Recovery Circuit for the ATLAS/CMS HL-LHC Pixel Front End Chip in 65-nm CMOS Technology ," in *Topical Workshop on Electronics for Particle Physics TWEPP 2019*, vol. 2, p. 6, 2019.

170. "MAG-PLL 1 MHz to 3 GHz Synthesizer." *MAGICS Technologies*, [Online] url: https://www.magics.tech/technologies/pll1001/, Accessed March 6, 2023.